Risk Management in an Uncertain World

Strategies for Crisis Management

Risk Management in an Uncertain World

Strategies for Crisis Management

Edited by
Bill Sharon

BLOOMSBURY

Copyright © Bloomsbury Information Ltd, 2012

Chapter pp. 77–84 copyright © Jonathan Silberstein-Loeb
Chapter pp. 85–93 copyright © Daniel Diermeier
Chapter pp. 95–107 copyright © Magnus Carter
Chapter pp. 109–117 copyright © Jon White

First published in 2012 by

Bloomsbury Information Ltd
50 Bedford Square
London
WC1B 3DP
United Kingdom

Bloomsbury Publishing, London, Berlin, New York, and Sydney
www.bloomsbury.com

All rights reserved; no part of this publication may be reproduced, stored in a retrieval system, or transmitted by any means, electronic, mechanical, photocopying, or otherwise, without the prior written permission of the publisher.

The information contained in this book is for general information purposes only. It does not constitute investment, financial, legal, or other advice, and should not be relied upon as such. No representation or warranty, express or implied, is made as to the accuracy or completeness of the contents. The publisher and the authors disclaim any warranty or liability for actions taken (or not taken) on the basis of information contained herein.

The views and opinions of the publisher may not necessarily coincide with some of the views and opinions expressed in this book, which are entirely those of the authors. No endorsement of them by the publisher should be inferred.

Every reasonable effort has been made to trace copyright holders of material reproduced in this book, but if any have been inadvertently overlooked then the publisher would be glad to hear from them.

A CIP record for this book is available from the British Library.

Standard edition	*Middle East edition*	*E-book edition*
ISBN-10: 1-84930-045-3	ISBN-10: 1-84930-052-6	ISBN-10: 1-84930-061-5
ISBN-13: 978-1-84930-045-2	ISBN-13: 978-1-84930-052-0	ISBN-13: 978-1-84930-061-2

Project Director: Conrad Gardner
Project Manager: Ben Hickling
Commissioning Editor: Lizzy Kingston

Typeset by Marsh Typesetting, West Sussex, UK
Printed and bound by CPI Group (UK) Ltd, Croydon, CR0 4YY

Contents

Introduction by Bill Sharon — vii
Contributors — ix

Risk in the Operational Disciplines

The Missing Metrics: Managing the Cost and Risk of Complexity
by John L. Mariotti — 3
Managing Operational Risks Using an All-Hazards Approach
by Mark Abkowitz — 11
Human Risk: How Effective Strategic Risk Management Can Identify Rogues
by Thomas McKaig — 17
Countering Supply Chain Risk by Vinod Lall — 23
Building Potential Catastrophe Management into a Strategic Risk Framework
by Duncan Martin — 29
Business Continuity Management: How to Prepare for the Worst
by Andrew Hiles — 37
Risk Management: Beyond Compliance by Bill Sharon — 45
How to Better Manage Your Financial Supply Chain by Juergen Bernd Weiss — 53
Managing Interest Rate Risk by Will Spinney — 59

Risk to an Organization's Reputation

Understanding Reputation Risk and Its Importance by Jenny Rayner — 69
Managing Reputational Risk: Business Journalism
by Jonathan Silberstein-Loeb — 77
The Cost of Reputation: The Impact of Events on a Company's Financial Performance by Daniel Diermeier — 85
Managing Your Reputation through Crisis: Opportunity or Threat?
by Magnus Carter — 95
Crisis Management and Strategies for Dealing with Crisis by Jon White — 109
Fraud: Minimizing the Impact on Corporate Image by Tim Johnson — 119

Risk in the External Environment

Measuring Company Exposure to Country Risk by Aswath Damodaran — 129
The Impact of Climate Change on Business by Graham Dawson — 135
Political Risk: Countering the Impact on Your Business by Ian Bremmer — 143

Introduction

The world is an increasingly uncertain place. The ability to predict events is becoming more difficult. Change is happening more quickly, facilitated by global social networking, with resultant political and social shifts. Governments fall, markets are volatile, and managing risk in this atmosphere is problematic. We are being forced to understand that what once seemed like a quantitative process, with defined outcomes, is actually more fluid; adaptation is the hallmark of effective risk management. Multiple scenarios can be planned for but, more often than not, organizations are surprised by situations that they had not anticipated.

We are discovering that compliance with laws and regulations, while an essential and necessary activity, offers little protection from the global increase in uncertainty. It is becoming clear that to act in the interests of their shareholders, organizations need to focus not only on what they want to do, but also on the context of those actions. The manner in which business is done separates the organizations that are resilient enough to weather the uncertainty of a rapidly changing environment from those who singularly focus on quarterly results.

The risk management profession has been influenced by a series of scandals and the resultant legislation (FDICIA, Sarbanes–Oxley, and Dodd–Frank in the United States) along with global protocols (Basel I, II, and III) designed to establish systemic stability. But obeying the law is not sufficient to guide a company through an unexpected event. We find that cultural values and the ability to respond and adjust when dealing with the unanticipated are the attributes that allow a company to not only survive but potentially to thrive in these circumstances.

Risk management is becoming a communications vehicle within organizations. The context for evaluating risk is determined by what a company wants to achieve rather than an objective set of rules with which it must comply. To that extent, our profession has become much more difficult; we must understand the culture and strategy of the companies we work for as well as their capabilities. The marriage of those elements means that risk is understood subjectively. We have become aware that the only certainty is uncertainty.

This book has assembled a group of experts from a wide variety of risk management disciplines. Although there is a view that managing uncertainty is primarily concerned with avoiding negative events, these authors focus more on what needs to go right than what might go wrong. To that end, we are seeing a shift in how risk is perceived. No longer can the risk manager promise that every negative event can be mitigated. Rather, the profession is increasingly looking at the approaches that will put a company in a position to respond effectively when a crisis occurs.

Collected under the heading "Risk in the Operational Disciplines," the first group of chapters deals with operational aspects of risk management. This encompasses everything from human resources, to supply chains, to the management of financial resources. As with any group of experts, you will find some views that contrast with others. Finding the right balance in these approaches is situational; companies have different cultures and operate in different environments. The aim here is to provide a range of opinion that will assist the reader in determining their own mix of strategies.

In the second group of chapters, "Risk to an Organization's Reputation," we deal with the issue of reputation risk. Many categories of risk can be quantified, but an organization's reputation is measured by its stakeholders' and consumers' attitudes and feelings about how it is behaving. As many of these authors point out, in a world where the unexpected is occurring with increasing frequency, the ability to manage reputation risk lies in the ethical framework within a company. It is clear that we are not just looking at downside risk here but rather awareness within an organization of opportunities to enhance its reputation as well. Risk management in this area lies in applying those ethical standards to a crisis situation. As these authors point out, the perception of how an organization handles a crisis is often more important than the initial impact of the crisis.

Our last group of chapters, "Risk in the External Environment," addresses risks that are beyond the capacity of an organization to control or influence. Climate change and political instability require an organization to be in a position to react swiftly to changing circumstances. Again we find that the theme of adaptability runs through these articles as the authors lay out the parameters of what an organization needs to contemplate when operating in multiple countries.

At the end of each article is a bibliography that will provide the reader with additional sources of further reading on the management of uncertainty. There is also biographical information for each author immediately following this introduction. Given the evolving nature of our profession it would be useful for us all to participate in an ongoing dialog; I would encourage you to reach out to these authors with your comments and questions.

Bill Sharon, Editor
CEO/Founder, SORMS, Inc.

Contributors

Mark Abkowitz is professor of civil and environmental engineering at Vanderbilt University and specializes in managing the risks associated with accidents, intentional acts, and natural disasters. He has a specific interest in the safety and security of hazardous materials, infrastructure adaptation to climate change, and in risk assessment using advanced information technologies. Dr Abkowitz has appeared on National Public Radio, Fox National News, and CNBC, discussing various risk management topics. From 2002 to 2011, he served as a Presidential appointee on the US Nuclear Waste Technical Review Board.

Ian Bremmer is the founder and president of Eurasia Group, a leading global political risk research and consulting firm. He created Wall Street's first global political risk index, and has authored several books, including *Every Nation for Itself: Winners and Losers in a G-Zero World* (2012), on risks and opportunities in a world without global leadership, and *The J Curve: A New Way to Understand Why Nations Rise and Fall*, which *The Economist* named one of the best books of 2006. He contributes to the *Financial Times* A-List, Reuters.com, and ForeignPolicy.com. Bremmer has a PhD in political science from Stanford University (1994), and he teaches at Columbia University. His analysis focuses on global macro political trends and emerging markets.

Magnus Carter is chairman of Mentor Communications Consultancy, which he founded in 1998. He lectures and coaches in media handling, reputation management, and issue and crisis management. His clients include NHS Blood and Transplant, the Office for National Statistics, Rolls-Royce, KPMG, and the Universities of Bedfordshire, Bristol, Sheffield, Southampton, and the West of England. He is an associate consultant of the Bristol Business School, a visiting lecturer at Ashridge Business School, and an approved trainer for the Chartered Institute of Public Relations. Carter began his career as a newspaper reporter and later gained more than 20 years' experience in radio and TV news, working with the BBC and commercial companies.

Aswath Damodaran is a professor of finance at the Stern School of Business at New York University, where he teaches corporate finance and equity valuation. He also teaches on the TRIUM Global Executive MBA program, an alliance of NYU Stern, the London School of Economics, and HEC School of Management. Professor Damodaran is best known as author of several widely used academic and practitioner texts on valuation, corporate finance, and investment management. He is also widely published in leading finance journals, including the *Journal of Financial and Quantitative Analysis,* the *Journal of Finance,* the *Journal of Financial Economics,* and the *Review of Financial Studies.*

Graham Dawson studied philosophy, politics, and economics at University College, Oxford, and holds a PhD in philosophy from the Keele University. He is the author of *Inflation and Unemployment: Causes, Consequences and Cures*, and of articles in journals including *Philosophy and Economics, Risk, Decision and Policy, Philosophy,* the *Review of Austrian Economics,* and *Economic Affairs.* He recently retired from the post of senior lecturer in economics at the Open University and is currently visiting fellow at the Max Beloff Centre for the Study of Liberty at the University of Buckingham.

Daniel Diermeier is the IBM professor of regulation and competitive practice, a professor of managerial economics and decision sciences at the Kellogg School of Management, and a professor of political science at the Weinberg College of Arts and Sciences, all at Northwestern University. He is director of the Ford Motor Company Center for Global Citizenship and co-creator of the CEO Perspective Program (Kellogg's most senior executive education program). Professor Diermeier's work focuses on reputation management, political and regulatory risk, crisis management, and integrated strategy. He was named Kellogg Professor of the Year (2001) and received the prestigious Faculty Pioneer Award from the Aspen Institute (2007). In December 2004 he was appointed to the Management Board of the FBI.

Andrew Hiles is founding director of Kingswell International, consultants and trainers in crisis, reputation, risk, continuity, and service management. He has conducted projects in some 60 countries. Kingswell International can be found at www.kingswell.net. He was founder and, for some 15 years, chairman of the first international user group for business continuity professionals, and founding director of the Business Continuity Institute and the World Food Safety Organisation. He has contributed to international standards and is the author of numerous books. He edited, and is the main contributor to, *The Definitive Handbook of Business Continuity Management*. Hiles has delivered more than 500 public and in-company workshops and training courses internationally and broadcasts on television, radio, webinars, and podcasts.

Tim Johnson is chief operating officer of Regester Larkin and oversees its international operations. He advises FTSE 100 and Fortune 500 companies and high-profile public organizations on how to earn, maintain, and expand their licenses to operate. Risk management and organizational resilience is Tim's specific area of expertise. He advises clients on the procedural, behavioral, cultural, leadership, and communication aspects of crisis preparedness and reputation management. He has supported clients facing allegations of fraud and senior management malpractice, industrial accidents leading to multiple fatalities, high-profile litigation, sensitive asset divestments, controversial market withdrawals, organizational restructuring, and supply chain failures.

Vinod Lall is a professor in the School of Business at Minnesota State University Moorhead, teaching supply chain management, operations management, management science, project management, and management information systems. Lall has developed and taught online and face-to-face graduate courses at business schools in Bulgaria, Ecuador, India, Thailand, and the United States. He has published numerous papers in peer-reviewed journals and presented at conferences. He is a certified supply chain professional (CSCP) by APICS, the American Association for Operations Management, and is the vice president of education for the Red River Valley chapter of APICS, leading certification training for a number of regional manufacturing and service organizations.

John L. Mariotti is the president, CEO, and founder of the Enterprise Group, a coalition of time-shared executive advisers. Mariotti is a director on corporate boards, including World Kitchen, LLC. Previously he was president of Rubbermaid Office Products Group (1992–94), leading nine divisions spanning four continents, and president of Huffy Bicycles (1982–92), the world's largest bicycle company in that era. Mariotti has written nine business books, a novel, thousands of articles, columns, and blog posts, and is a highly regarded keynote speaker. His book *The Complexity Crisis* received awards as one of 2008's best business books. He was previously a contributing editor for *Industry Week* magazine and he currently writes for Forbes.com. He can be found at www.mariotti.net

Duncan Martin is a partner and managing director in the risk management practice at the Boston Consulting Group (BCG), based in London. Prior to joining BCG, he was the head of Wholesale Credit Risk Analytics at the Royal Bank of Scotland in London, the director of Strategic Risk Management at Dresdner Kleinwort, and a senior manager at Oliver Wyman & Company. Martin was educated at Cambridge University and the Wharton School of the University of Pennsylvania. He is the author of the book *Managing Risk in Extreme Environments* (2008).

Thomas McKaig is a widely recognized Canadian author with 30 years of international business experience in 40+ countries. He owns Thomas McKaig International, Inc., found at www.tm-int.com. He speaks internationally on quality management and international trade, and is an adjunct professor, teaching Global Business Today in the Executive MBA program at the University of Guelph. His most recent book is *Global Business Today* (3rd ed, 2011). He has served as executive in residence at the University of Tennessee and Universidad de Montevideo, and was worldwide strategic marketing adviser to the United States Treasury Department Bureau of the US Mint's Gold Eagle Bullion coin program.

Jenny Rayner is director and principal consultant at Abbey Consulting, which she established in 1999 to provide consultancy and training on the positive management of risk to improve business performance and protect and enhance reputation. Prior to this, her wide-ranging career spanned more than 20 years with ICI and Zeneca in a variety of sales, marketing, purchasing, logistics, supply chain, and general business management roles, and latterly she was a chief internal auditor with ICI. Rayner also trains, writes, and lectures on risk management, corporate governance, internal audit, corporate social responsibility, and reputation.

Bill Sharon has been conducting seminars, workshops, and consulting assignments in the area of risk management for the past 14 years. He has 30 years' experience in the financial services and marketing/communications industry in a variety of C-level positions and consultancies. He has been featured in numerous industry magazines (*Intelligent Risk, CIO Magazine, Business Finance*, and *Business Credit Magazine*) and has authored numerous articles. Bill holds a clinical degree, and for the first 10 years of his professional life he worked with adolescents—an experience that taught him the very difficult skill of how to listen. His website, Strategic Operational Risk Management Solutions, can be found at www.sorms.com.

Jonathan Silberstein-Loeb is a research fellow at the Oxford University Centre for Corporate Reputation. He was an Alfred Chandler Jr traveling fellow at Harvard Business School, a fellow at the Newberry Library, Chicago, a fellow of the Lilly Library at the University of Indiana, Bloomington, a Deutscher Akademischer Austauschdienst fellow in Berlin, and a Fulbright Fellow in Kyoto, Japan. He received his doctorate from the University of Cambridge.

Will Spinney joined the treasury department at Johnson Matthey plc after a brief career in the Royal Navy, and took the first ever Association of Corporate Treasurers (ACT) corporate treasury exams in 1985. He has been a practicing treasurer now for 25 years, working for several companies that have included most recently Eaton Corporation and Invensys plc, where his experience ranged from risk management, cash management, and extensive refinancings to pension investment strategies. He has been a speaker at several ACT conferences and has been involved in education and training programs with the ACT for several years, both writing resources and as a member of the MCT examination board.

Juergen Bernd Weiss worked for almost 11 years at SAP AG in the SAP ERP (enterprise resource planning) financials area. In his last position he was consulting director for financial supply chain management and corporate performance management. He also worked as director of application solution management ERP and held global responsibility for financial supply chain management, particularly for SAP solutions in the areas of electronic bill presentment and payment, dispute management, credit management, in-house cash, and customer and vendor accounting. Prior to joining SAP in 1997 he worked for Westdeutsche Landesbank in Düsseldorf, Germany. He has degrees in economics and business.

Jon White is a consultant specializing in the application of psychological thinking to the problems of organizational communication, working internationally for clients such as the European Commission and Shell. He is associated with Henley Business School, Cardiff University, the University of Central Lancashire, and universities in Germany and Switzerland, for teaching, research, and special projects. He has written books and articles on public affairs, public relations, and corporate communications practice, and management case studies for teaching purposes on organizations such as Dunhill, Lloyds of London, AEA Technology, Diageo, and the South African company Barloworld. He is a fellow of the UK's Chartered Institute of Public Relations.

Risk in the Operational Disciplines

The Missing Metrics: Managing the Cost and Risk of Complexity

by John L. Mariotti
President and CEO of The Enterprise Group, Powell, Ohio, USA

This Chapter Covers

- Despite the best efforts in many areas, the accounting and finance systems currently in use overlook the costs of complexity—it's all talk and too little action.
- Complexity costs are hidden, buried in the accounts of a company until the period-end statements reflect the adverse effects on profitability.
- The time to recognize that these costs exist is now, after which they can be identified and new metrics found or developed to take the place of those that have been missing for far too long.

Introduction

Accounting systems have come a long way in the past decades. Activity-based costing revealed where costs were being incurred and what was driving them. The blizzard of regulations following debacles involving Enron, WorldCom, etc., led to the passage of the Sarbanes–Oxley Act. The most recent financial crisis spawned the Dodd–Frank Wall Street Reform and Consumer Protection Act (2010), which will undoubtedly lead to many more new regulations (in the United States).

These burdensome new laws impose some necessary disciplines on finance and accounting but fail to deal with a huge unmeasured and unmanaged area—the costs of complexity. When I began studying this area in earnest about a decade ago, I discovered just how far-reaching the negative impact of complexity has grown, and how much it has gone unnoticed. Certainly there is discussion of complexity, but following all the talk there is very little organized action.

Back in 2001, Oracle CEO Larry Ellison described a "war on complexity" in computer software. There were simply too many systems that were not integrated, and others that were very difficult to integrate. This fragmentation of systems caused huge complexity, duplication of effort, and waste—which Ellison's Oracle Corporation hoped to solve.

In 2006–08 there was another flurry of reports and articles from major consultancies (Bain, McKinsey) and universities (Wharton, Harvard). More recently, IBM's May 2010 report described a comprehensive survey of 1,500 CEOs, which confirmed the breadth, depth, and magnitude of the risks of complexity and how dangerous it is to do business around the world. And yet words have not been converted into action effectively, and complexity management continues to be situational and far from fully effective. A great opportunity still awaits global business.

Variety Can Add Value—If Managed Properly
There are clearly instances where complexity, properly managed, can be a source of great competitive advantage. In these cases, the organization's structure, systems, and processes must be carefully designed to minimize transaction costs and make complexity manageable.

One notable success in the use of complexity for competitive advantage is the web retailer Amazon, whose breadth of offering is extensive and growing, thus making it a "one-stop shopping" site for millions. Amazon's distribution system, however, is always at risk of being overwhelmed by complexity, even as its front-end systems handle the huge variety of goods which it sells seamlessly.

Similarly, US sandwich seller Subway assembles sandwiches to order from 30–40 containers of meat, cheese, and vegetables, using just a dozen varieties of breads and wraps. Thus it can make to order millions of sandwiches (and salads) with minimal waste and great flexibility.

There are many other examples like these two. All depend on the right systemic design to keep complexity from growing out of control, causing harmful, costly waste and inefficiency.

Complexity Costs Are Hidden
When I first researched why the costs of complexity remained unmeasured in so many companies, I discovered that it was because these costs are, by their nature, hidden by conventional accounting systems. To bring the problem into perspective, consider how complexity occurs and what kinds of waste result. It will become apparent how financial systems simply overlook complexity's costs—until the end-of-period reporting shows the detrimental effects and true costs.

There is no doubt that complexity's effects are readily apparent in monthly, quarterly, and year-end results, where they reduce income and impact the balance sheet adversely. Unfortunately, this is too often the only time and place where they are visible. Even then, there's no indication of how or where these costs were incurred, or how to manage and minimize them.

Seeking High Growth in Low-Growth Markets
So much of the complexity that goes unmeasured and unmanaged is created, with the best of intentions, in search of revenue growth. Many developed countries (the United States, Europe, Japan, etc.) are growing very slowly, in population and economically. When companies seek growth in such mature markets, they resort to proliferation, which leads to complexity. The gain in revenue is redistributed across a broader range of products and services, with only modest increases in total revenue. The many resulting new products, customers, markets, and suppliers add much more in complexity costs than in profit. As rapidly growing economies like China slow, even slightly, complexity costs will start to impact them as well.

Mergers and acquisitions are another source of complexity. If either of the two combined companies is burdened with complexity (most are), this will transfer

The Missing Metrics: Managing the Cost and Risk of Complexity

to the merger. If both are thus burdened, real trouble is likely. Simply combining the DNA of two companies is a daunting task without struggling under a burden of being "infected" with the complexities of two different "strains." There are issues of product and customer overlap, organizational and/or facility redundancy, and inevitable information systems redundancies. When these are combined with cultural conflicts that must be sorted out, the problems become almost insurmountable. This is one of the main reasons why mergers seldom lead to long-term growth in shareholder value.

Less developed countries typically grow at much higher rates (China, India, Brazil, etc.). Emerging consumer societies with favorable balances of trade fuel their economic growth. There's a different complexity problem here: most of these countries save more and spend less—both as consumers and as governments. Further, these countries are less familiar to sellers who operate in developed countries, and therefore marketing and operating mistakes are made. These mistakes also lead to proliferation, often due to errors in targeting, product configuration, branding, and/or how to serve the targeted markets and customers.

Profits Are Proportional to Revenues; Costs Are Proportional to Transactions

Thus, either approach to growth adds to complexity, but for different reasons and in different ways. Profits are derived from increased revenues, but costs are incurred by increased transactions. Therein lies the root of the complexity problems. A few simple reports can expose the problem, but more sophisticated solutions are needed later. Starting with the simpler metrics is advisable (I will say more on "simplicity" later).

First, calculate sales per customer, per product, per location, etc., and track the trends. They are typically declining, indicating more transactions for less revenue.

Next, sort the annual sales, profits, etc., for customers and products, in descending order of value, and compute a cumulative column. Now look at the bottom of the list. There is always page after page of "losers" with few sales and low or negative profits. These are the candidates for a major "house-cleaning."

Few accounting systems calculate a couple of simple, yet important, measures. What is the cost to process a customer order from "end to end"—from receipt of the order until the payment is in the bank? Few, if any, companies know the answer to this question. One US study, performed by Sterling Commerce a few years ago, calculated it to be approximately $50. Consider the following quick calculation to show how complexity adds cost and waste.

Many companies make about 5% net profit (after tax) on sales revenue. That means that $20 of sales will generate $1 of net profit. If processing an order costs $50, they need a $1,000 order to earn the equivalent of what it costs to process the order. If that type of customer orders every week, $50,000 worth of annual sales barely generates net profit equal to the cost of processing the orders.

Risk Management in an Uncertain World

This dramatically illustrates how customer orders that are small and frequent can add complexity cost, and yet this cost remains unmeasured and undetected as a drain on profit. A similar comparison could be made for the cost of processing purchase orders, or the expense of setting up/maintaining documentation for a product or service. Nowhere are these costs measured in this way, and thus they remain unmanaged. Most companies have departments that perform these functions. Totaling those departmental expenses and dividing by the total number of orders processed will yield an adequate approximation of the cost of processing each order. Yet few companies do this calculation—or consider its impact.

Complexity costs are also insidious because most of them are hidden in "catch-all" accounts such as variances, allowances and deductions, premium freight, need for overtime labor, scrap and rework, closeout pricing, and so forth. Extra effort is needed to reveal the origin of such entries (more on that later).

Consider a simple example of how easily complexity can occur and grow.

A Simple Example: One White Coffee Mug

Imagine a coffee mug in one style, color, size, and type of packaging; sourced from one supplier, packaged and stocked in one location; and offered for sale to one customer. If the mug's total landed "standard cost" is $1 and it sells for $2, this yields a 50% gross profit margin.

Because the mug is a successful product, the company expands the line to four styles, four colors, two sizes, and two packaging options. There are now 64 different variations, which leads to complexity (and errors) in forecasting, buying, controlling, and managing inventory, etc. The "standard cost," however, is still the same as before: cost = $1, price = $2, and gross profit margin = 50%. *But something is wrong.* Intuitively, you know that there are complexity costs that the old metrics don't capture—at least, which are not included in the "standard cost" and gross profit margin calculation. The true profitability is not nearly the same as before—it is lower, maybe much lower.

Success in revenue growth leads to expansion of the product line again: two suppliers, packaging and inventory in three locations, and selling into three more countries (or markets). Assuming no difference in purchase cost or productivity, the standard cost, price, and gross profit margin remain the same. But now the number of combinations and permutations has grown to over one thousand. Many different marketing materials are needed, the risk of purchasing/forecast errors grows with demand volatility, and so on. Now the true profitability is clearly lowered again—complexity has struck.

Add color mixes and assortments of product that vary by market, customer, production plant, distribution center, and country, and the warehouses fill with products in the wrong colors or styles, wrong package sizes, etc. Something must be done with these oddments, so they are repacked (cost variance) and sold at discounts (price variance), and replacements are flown in (huge freight expense variances) to meet customer service needs. More profit disappears into those

The Missing Metrics: Managing the Cost and Risk of Complexity

"catch-all" accounts. The accounting standard gross margin has remained essentially unchanged.

Complexity creates increases in overhead and administrative expenses; in the reserves needed to cover inventory obsolescence; and/or incurs additional labor to rework, repack, and remark inventory. Few of these costs impact the standard cost and the standard gross margin. Thus, the product still appears to be nicely profitable. The complexity costs remain hidden in undifferentiated accounts—or result in "nonrecurring charges," which, mysteriously, seem to "recur" from time to time. At the end of accounting periods, the true costs hit with full impact, in many cases wiping out all profit.

A Complexity Crisis Calls for Metrics

I call this sequence of events "a complexity crisis." The finance and accounting metrics, intended to help track the results of the company, do so—eventually. Unfortunately, the waste from complexity remains unmanaged; the missing metrics do not reveal the problems. They are seen only after the fact. Complexity strikes like a robber. A crime has been committed; the money is gone. Clues to the crime are few, and the perpetrators plead innocence and good intentions. Only a knowledgeable accountant, with help from supply chain or marketing staff, can unearth the clues and track the loss of money back to its root causes.

The solution for this is evident: to devise and implement the "missing metrics." Many are easy to create; some are already in use. Others will require whole new initiatives (described later). If new metrics were in place and tracked regularly, such losses would be found much sooner. Then corrective actions could be started sooner as well. Major public accounting companies could help by sanctioning such metrics, to provide some uniformity. Unfortunately, thus far they have been unresponsive to those needs. Perhaps they are too busy dealing with new governmental regulations and compliance to watch out for the client company's profits.

Typical Missing Metrics

Typical missing metrics are:

- sales per product stock-keeping unit (SKU);
- sales per product category;
- sales per customer;
- sales per location;
- sales per employee (hourly, including full-time equivalent, salaried, and total);
- sales per part number (components, materials, work in process, and finished goods (FG)).
- gross profit per product SKU;
- gross profit per product category;
- gross profit per customer;
- gross profit per location.
- purchases per vendor;
- purchases per commodity type;

Risk Management in an Uncertain World

- production (output value) per person-hour (or equivalent measure of labor input);
- total number of SKUs by division or business unit and company total;
- number of SKUs added and dropped during the last time period (quarterly, semiannually, or annually);
- cost to process a customer order (end to end);
- cost to process a purchase order (end to end);
- cost to set up and maintain a product SKU;
- cost to serve by customer (including freight, handling, and order processing costs).

Within "catchall accounts" like deductions and allowances, variances, and writeoffs for obsolescence, add subcategories to segregate entries by major customers (or groups), products (or categories), and locations (divisions):

- expenses per product line or category;
- expenses per customer, and/or by customer type/category;
- expenses per location;
- percentage of sales per product line or category;
- percentage of sales per customer, and customer type or category.

Plus a Totally New Metric—The Complexity Factor (CF)—And New Tools

Obviously, there is a common overall purpose among these metrics. Remember that the objective of new metrics is to reveal where the costs of complexity are hiding and are wasting time and money. Choose among those that measure similar complexity-related outcomes. Finally, an overall complexity factor can be calculated by the following formula (where "locations" are meaningful facilities and "countries" are places where legal entities exist):

(No. of suppliers + No. of customers + No. of employees) × No. of FG SKUs × No. of markets served ×

No. of locations × No. of countries ÷ Total annual sales revenue

where total annual sales revenue can be expressed in the company's currency of choice. The resultant number provides a "benchmark," called a complexity factor (CF), for the business (or subunit) whose data were used to calculate it. Obviously, a CF can be calculated for each business unit, division, geographical unit, etc., *and* for the entire company.

This may seem like a large number of new metrics, but the data to compile them should already exist. All parts of a company may not need all of the metrics. Different parts of the business should use metrics (including the CF) that are relevant to their activities.

Some much simpler solutions have been developed in two different continents: by www.simpler.com in the United Kingdom, and by www.simplerbusiness.com in Australia. Each offers a range of solutions that focus on simplicity instead of complexity. Whichever solution you choose to manage complexity will use largely

The Missing Metrics: Managing the Cost and Risk of Complexity

similar approaches and metrics: find where the complexity resides and how much it costs, and then drive it out and manage it. Complexity is like the weeds in a garden. Removing them once is not sufficient. They come back. Like weeds, complexity must be constantly measured and managed.

When more sophisticated, analytical solutions are needed, new ones have been developed by Emcien (www.emcien.com), which analyzes and optimizes complex patterns/assortments, and by Ontonix (www.ontonix.com), which also measures risk. As complexity grows, an organization can be overwhelmed by it and lose control of the company. This is a catastrophic failure and must be avoided at all costs. The OntoNet system by Ontonix analyzes complexity in terms of risk and the fragility of an entity, and can be applied to a wide range of situations. This kind of solution is the ultimate "insurance policy" missing metric for complexity risk.

What Gets Measured, Gets Managed; What Doesn't, Doesn't
The mere presence of metrics does not mean that management will do anything differently. On the other hand, the absence of metrics virtually assures that nothing will be done. The old line, "What gets measured, gets managed," is true. The opposite, "If you can't (or don't) measure it, you can't (or won't) manage it," is also likely to be true. Talk and studies do not solve problems; only action does.

Measurement alone also doesn't solve any problems either. It merely points to the nature of those problems. To manage complexity requires a series of steps from simpler to more sophisticated.

First, use Pareto's principle (the 80–20 rule). Sort products and customers in descending order of annual revenues and profits, and carefully analyze the bottom of the list. Most are "losers," with a few strategically important "potential winners" scattered about. Getting rid of the losers is imperative.

Upgrading a few "losers" into "winners" (the top 20%) is possible, but for most it is impractical. In the middle group careful analysis helps to identify potential winners for upgrading and to spot imminent losers for downgrading and removal.

When more powerful tools are needed, the newly devised, more powerful tools and systems (Emcien, Ontonix) can help greatly in the sorting, selection, and optimization of complex situations.

The Time for New Metrics Is Now
Now is the time for boards of directors, senior management, and accounting and finance organizations around the globe to recognize the huge cost of complexity and how poorly measured and managed it is. The waste of time and money due to "missing metrics" and the failure to track and manage complexity is immense, costing companies around the globe billions in profits. Now is the time to stop that waste in its tracks—by installing those "missing metrics" and then acting on the new information and knowledge which they provide.

More Info

Books:

George, Michael L., and Stephen A. Wilson. *Conquering Complexity in Your Business*. New York: McGraw-Hill, 2004.

Mariotti, John. *The Complexity Crisis: Why Too Many Products, Markets, and Customers Are Crippling Your Company—And What to Do About It*. Avon, MA: Adams Media, 2008.

Articles:

Berlind, David. "Oracle: Misquoted, misunderstood." *ZDnet Tech Update* (September 6, 2001). Online at: tinyurl.com/3usfr6a

Gottfredson, Mark, and Keith Aspinall. "Innovation versus complexity: What is too much of a good thing?" *Harvard Business Review* 83:11 (November 2005): 62–71. Online at: tinyurl.com/3lkanaw

Heywood, Suzanne, Jessica Spungin, and David Turnbull. "Cracking the complexity code." *McKinsey Quarterly* (May 2007): 85–95. Online at: tinyurl.com/cejypo

Reports:

A.T. Kearney. "Waging war on complexity: How to master the matrix organizational structure." A.T. Kearney, February 2002.
Online at: www.managementplace.com/fr/atk/waron.pdf

George Group and Knowledge@Wharton. "Unraveling complexity in products and services." Special report, 2006. Online at: tinyurl.com/3u4lrke

IBM Global Business Services. "Capitalizing on complexity: Insights from the Global Chief Executive Officer Study." IBM Corporation, May 2010.
Online at: tinyurl.com/ca38wmn

Websites:

Emcien, optimization tools for complex solutions: www.emcien.com

Ontonix, complexity and risk management tools: www.ontonix.com

Simpler (United Kingdom): www.simpler.com

Simpler Business (Australia): www.simplerbusiness.com

Managing Operational Risks Using an All-Hazards Approach

by Mark Abkowitz
Vanderbilt School of Engineering, Nashville, Tennessee, USA

This Chapter Covers

- Operational risk management (ORM) enables an enterprise to understand, prioritize, and control risks that threaten its well-being and the livelihood of its partners.
- Although traditionally stove-piped within an organization, different operational risks share many common elements, providing an opportunity to consolidate ORM into a single all-hazards approach, one that is holistic and systematic.
- The key to effective ORM is to recognize and mitigate those risk factors that erode our margin of safety, so allowing situations to spiral out of control.
- A key first step is for an organization to perform an ORM physical, enabling the identification of reasonably foreseeable risks, benchmarking the current status of the ORM program, revealing gaps where the organization is vulnerable, and developing cost-effective strategies to address these gaps.
- Based on recent historical events and changing conditions in our world, bringing ORM to the forefront of an organization is more important now than ever before.

Operational Risk Management: A Definition and a Strategy

For the purpose of this discussion, operational risk management (ORM) is considered to be the policies, methods, practices, and institutional culture that enable an enterprise to understand, prioritize, and control risks that threaten the well-being of the organization, its business partners, communities in which it operates, and society at large.

The cost of *poor* operational risk management can be excessive, considering that the occurrence of undesirable events can lead to fatalities and injuries; property loss; business interruption; clean-up, remediation and disposal; fines and penalties; future inspections; new regulations; long-term human health effects; environmental degradation; damaged investor, insurer, supplier, and customer relations; and loss of public confidence. By contrast, the cost of *good* operational risk management may be limited to investment in risk management benchmarking and needs assessment; resources allocated to control high-priority risks; and ongoing costs associated with ORM performance monitoring and evaluation.

The Need for an All-Hazards Approach

In many organizations, the approach to dealing with operational risks is stove-piped, with different entities having responsibility for different hazards. For example, environmental health and safety worries about toxicity exposure, legal is concerned with liability, human resources focuses on occupational health, executive management has its eye on business continuity, risk management addresses insurance, and research and development cares about design failure. Each group has its own priorities, separate resources are used to address each problem, and there is limited coordination.

Risk Management in an Uncertain World

Yet, while each threat may seem quite different, when one takes a closer look at how these events evolve, there is remarkable similarity; that is, a pattern or "recipe" for disaster emerges. This situation begs for the adoption of a single "all-hazards" ORM approach, a process that is holistic and systematic in nature.

Risk Factors

Within a recipe for disaster, each ingredient can be thought of as an underlying risk factor that erodes our margin of safety. Once this margin of safety is exceeded, the situation is liable to spiral out of control. Therefore, management control of risk factors is at the crux of an effective ORM program. In attempting to manage these risk factors within an organization, it is helpful to group them into the following categories:

Design and construction flaws: If there is a flaw in the design process and it is not discovered in time, the system is prone to failure. Even when the design is valid, problems can still arise if the materials used to fabricate the system components are faulty or the components are not assembled properly.

Deferred maintenance: It is human nature to choose to deal with problems at a later time, especially if the system is not actually malfunctioning. Unfortunately, decisions to defer maintenance often lead to the failure of a key system component before the repair can be made, causing a serious accident to occur.

Economic pressures: Organizations typically manage a limited budget. When these resources are too scarce or spending is not controlled adequately, pressure intensifies to implement strict cost-cutting measures. This can lead to shoddy workmanship, the purchase of inferior quality materials, elimination of the use of backup operating and safety equipment, or management ignoring problems that arise.

Schedule constraints: When a deadline has been imposed, and the activity has fallen behind schedule, pressure to make up ground can cause the responsible party to turn a blind eye to important details. This situation often leads to the elimination of critical tasks, personnel trying to accomplish tasks in parallel that should be done in sequence, or not pursuing certain considerations in sufficient depth to understand their impact on safety fully.

Inadequate training: Because of a lack of adequate training, individuals who are prone to make mistakes may be placed in positions of responsibility. This in turn can either initiate or intensify a crisis situation. When there are personnel shortages, individuals may be thrown into an important decision-making role while covering for others, performing a function for which they were not properly trained. Because individuals tend to forget what they were originally taught and since processes change over time and require new learning, lack of retraining can also be a problem.

Not following procedures: When engaged in a repetitive activity, complacency can set in, and individuals tend to drift away from following formal protocols. Consequently, they either neglect to perform certain steps or invent other ways to accomplish the same task, often not considering the possible safety hazards caused by their actions. Failing to follow procedures can create a hazardous situation, one that is exacerbated

Managing Operational Risks Using an All-Hazards Approach

by coworkers whose actions are based on assuming that those procedures are being followed.

Lack of planning and preparedness: Because of the luxury of time and the fact that a disastrous event may not have been experienced in recent memory, people tend to place a low priority on being adequately prepared for a crisis situation. All too often, little forethought is given to the variety of disaster scenarios that could reasonably occur and how to deal with them effectively. Even in circumstances where significant effort has been devoted to planning and preparedness, the product of this effort can be a written plan that is not practiced or updated, rendering it of little value when a calamity arises. Lack of planning and preparedness is one of the most common risk factors at play when something goes wrong.

Communication failure: Communication failures can occur at various stages, altering an outcome in different ways. When communication fails between members of the same organization, critical information is not shared, such as when one group decides to shut down a critical protection system for maintenance while another group is carrying out a dangerous experiment. Poor communication between organizations is also problematic. Finally, lack of communication with the public or the provision of inaccurate information can place people at risk either because they do not know the hazards they are facing, or because they are not properly advised on how to protect themselves. Along with lack of planning and preparedness, communication failure is the most common risk factor at play when something goes wrong.

Arrogance: Arrogance can rear its head in many forms, but usually appears as either the person in charge being driven to succeed for individual gain without sufficient regard for the safety of others, or an experienced individual who has become overconfident in his or her ability to deal with any problem that might present itself. In either form, arrogance can have serious repercussions.

Stifling political agendas: Government policies can have a powerful effect on the propensity for disasters. If these political agendas are hard-nosed, with little room for dialog and compromise, affected parties can feel that they have little recourse other than to resort to extreme and often hostile measures.

It is important to note that we, as humans, are involved in each and every one of these factors. While this implies that we contribute to the cause or impact of every disaster, it also means that we have an opportunity to control these factors more effectively to achieve a better future outcome.

Getting Started

A key first step is for your organization to have an ORM physical, essentially a comprehensive review of how operations are performed, what risks are present in performing these operations, and how these risks are presently being managed. This engages the organization in identifying "reasonably foreseeable" risks, benchmarking the current status of the existing ORM program, whether relatively new or fairly mature, identifying program gaps where the organization carries the greatest liability, and suggesting strategies and tactics that can be implemented to close these gaps.

Risk Management in an Uncertain World

Case Studies

ORM Failures and Successes

There are several historic events that bring the failures and successes of operational risk management into focus. How could the event have been prevented? What could have been done to mitigate the impacts? What management controls have been implemented since the event occurred? Could it happen again? These are all legitimate ORM questions that, through hindsight, allow us to learn from experience and apply these lessons to deploying more effective ORM in the future.

Hurricane Katrina

During August 2005, Hurricane Katrina slammed into the United States, hitting the coastal areas of Florida, Louisiana, and Mississippi. A combination of storm surge, wave action, and high winds resulted in the destruction of buildings and roads in the affected areas. The impact of Katrina on New Orleans was unusually severe; portions of the city were left under 20 feet of water due to failure of the earthen levees and floodwalls that had been constructed to safeguard the city from this type of event. Hurricane Katrina caused nearly 2,000 fatalities and an estimated economic loss of US$125 billion, in addition to displacing hundreds of thousands of people from their homes and workplaces. The destruction and loss of life in New Orleans, while initiated by the storm itself, cannot be attributed entirely to Katrina. Numerous failures of the city's flood protection system due to poor design and construction, deferred maintenance, and a lack of funding left New Orleans susceptible to a hurricane of Katrina's magnitude. As the city filled with water, the hurricane's effects were compounded by insufficient emergency planning and preparedness, and the inability of responders to communicate.

Alaska Pipeline and Denali Earthquake

A major earthquake struck the Alaska mainland on November 3, 2002, along the Denali fault, which passes directly under the Trans-Alaska Pipeline. Had the pipeline ruptured, it would have resulted in spillage of up to a million barrels of crude oil a day in an environmentally sensitive area. Yet not a drop of oil was released. This potential catastrophe was averted due to successful ORM in both the design of the pipeline system and the quality of the maintenance, surveillance, and emergency preparedness. The pipeline design team, using extensive field data, devised a system such that it could survive a major earthquake should one occur during the pipeline's projected 300-year operating period. As a result, a US$3 million upfront investment in geological studies and corresponding design considerations helped to prevent an environmental disaster that could easily have topped US$100 million in remediation costs. Concurrently, a comprehensive surveillance and maintenance system was implemented, capable of identifying problem locations in real time and dispatching crews accordingly. Moreover, emergency response was facilitated by a well-organized incident command system, contingency planning, and a training program.

Managing Operational Risks Using an All-Hazards Approach

Making It Happen

- Designate ORM as a core business practice within the organization by establishing the program at the vice-president level. The VP should be responsible for defining ORM policies and procedures, and for providing oversight of program activities.
- Organize an ORM committee, which reports to the VP, with membership that includes representatives from each element of the organization that has a designated ORM responsibility.
- Perform an ORM physical, and use it as a basis for defining program priorities, allocating resources, and implementing management control strategies.
- Monitor and evaluate ORM performance to determine whether program objectives are being met.
- Maintain ORM as a living process that is part of the culture of the organization.

Conclusion

We can ill afford not to recognize the new age of operational risk management, one based on a holistic and systematic approach to identifying reasonably foreseeable risks, establishing priorities, and adopting practical, achievable, and cost-effective control strategies. As history has taught us, we remain vulnerable to the occurrence of catastrophic events whose prevention or mitigation is within our control. Moreover, changing conditions in our world are posing new challenges that will require making tough risk-related choices. Adopting an all-hazards ORM approach does not mean that we will never suffer another tragedy. However, the prospect of that happening is less likely to occur once investments in prevention and mitigation have been made. The bottom line is that we can, and should, do much better at being a master rather than a victim of risk. All it takes is a more organized approach to managing the risks that affect our daily lives, coupled with a greater tolerance for unfortunate events that will sometimes occur no matter how hard we try to avoid or prevent them.

More Info

Books:
Abkowitz, Mark D. *Operational Risk Management: A Case Study Approach to Effective Planning and Response*. Hoboken, NJ: Wiley, 2008.
Garrick, B. John. *Quantifying and Controlling Catastrophic Risks*. San Diego, CA: Elsevier, 2008.

Websites:
Risk World: www.riskworld.com
Society for Risk Analysis (SRA): www.sra.org

Human Risk: How Effective Strategic Risk Management Can Identify Rogues

by Thomas McKaig

Thomas McKaig International Inc., Ontario, Canada

This Chapter Covers

- Corporations and high-level risk management are built around the people in organizations—and people are fallible.
- The need to evaluate human risk is clear: Stories abound of rogue employees in large and small organizations who have destroyed their entire firm.
- At the extreme, rogue firms, such as Enron, can destroy shareholder value and employees' lives.
- Building a quality-based organization helps to drive out rogues, but that's not the only way.
- Control measures need to be in place.
- Legal measures, the spotlight of publicity, and backing up corporate policies with firm action are all effective tools.

Introduction

Best practices in strategic risk management are intended to prevent weaknesses within corporations causing damage or even pulling down the firm. However, effective strategic risk management tools and techniques became harder to implement as business operations grow, become more complex, and operate in multiple locations. The controls that might have once been deemed acceptable in keeping employees within corporations on the same page begin to be less effective in cases of corporate restructurings that split businesses into smaller business units, and where employees are prodded into making deeper contributions to the bottom line.

Technology has not necessarily been a savior in this type of situation. Although technology has provided a platform for enhancing competitive advantage for business, it has also been a tool used by smart, capable, yet ill-intentioned employees to steal and distort overall results.

In the age of managerial cutbacks and increased workloads, a lot of things can happen that go unnoticed by overburdened managers. Interview techniques intended to keep rogues out of the workplace are—in spite of all the high-end questionnaires and intensive interview techniques that may be used—oftentimes ineffective, as potential employees are extremely savvy about modern interview techniques. Players in the job market are often familiar with the drill. Job hunters pass through many revolving interview doors, allowing them to hone their skills on how to dupe the interview process. Some interviewers may be incompetent or show poor judgment. HR departments are not foolproof, and it is only realistic to accept the fact that rogues in the workplace are here to stay. HR people will sometimes catch potential wrongdoers at the gatepost through psychological tests and other forms of due diligence involving intuition and criminal checks. But don't count on it.

Risk Management in an Uncertain World

Newspapers are full of stories about accountants who pad the books and give kickbacks to friends and family. Unhappy workers can damage product on the assembly line. A fired employee can show up at the workplace intent on payback for the injustice he or she feels they have suffered (in the United States this is called "going postal"). A multinational manager away from the watchful eyes of the home office can withhold information and deliver selective reports. Expense accounts can be padded. Goods can be pilfered from warehouses.

Given the current economic and political shocks, the last thing a company needs is to find itself in the news on account of the excessive creativity of one or more of its employees. Managers must face the fact that rogues will enter their organizations. So the question becomes: What can be done about it before the damage is done?

Keep in mind that human risk is about more than employees stealing from a firm; it can include individuals making unsound business decisions because nobody told them otherwise. Mistakes can be just as bad as deliberate fraud, as the following case shows.

Case Study

An Invitation to Rogue Employees

The example of a small Costa Rican bank serves to illustrate this point. At the height of the opening of Costa Rica's financial markets to foreign financial institutions in 1995 there was a rush to change operations practice. In the pre-free market era, Costa Rican banks could do as they pleased and were immune to punishment even when there were banking scandals and losses that were large for Costa Rica's fragile economy during the 1980s and 1990s. Old-style banks, accustomed to getting away with providing poor customer service and having lax internal controls, found that their business environment was changing with the pending legislative changes, set to open Costa Rica's financial markets to the world.

With poor leadership at the helm, and a lack of almost any strategic management initiative, employees were forced to take on new and undefined roles in their bank. Most of these were ill-suited to employees who were given inadequate training and guidance for their new tasks.

As part of rising to the challenge of this expected competition from foreign banks, and in light of the assumed effectiveness of recently ordered ATM machines, the bank we are considering decided that a lean and mean (and ill-informed) policy of rampant firing would be an acceptable cost-saving measure. Half of the bank's staff lost their jobs, and those who remained quickly became demoralized. The newly installed bank machines did not function properly. Friday afternoon payday waits grew to two hours from the already unacceptable 15–30 minutes.

Internal communications broke down. In place of the usual courteous conversations, vitriolic emails flew from one cubicle to the next—seeding the environment for "surprise actions" from a growing league of unhappy, overworked, and demoralized employees. With no controls in place, an inexperienced bank teller authorized a loan of US$1 million to a long-standing customer—based solely on the fact that the teller liked the man and felt that he could be trusted with the money. For a small bank with a net worth of US$37 million, this

How Effective Strategic Risk Management Can Identify Rogues

> inappropriate loan decision was the start of a string of poor management decisions that led to its implosion. Throughout this process the business culture undermined any attempts to implement benchmarking studies or best-practice management solutions. The "generous" employee was not fired and kept his duties with a severe reprimand. The future of the bank was sealed, and eventually it went down.

At the Extreme

At the extreme end of the spectrum, there is a widespread pattern of "pushing the boundaries" of everything from accounting rules to disclosure rules for public companies, lax internal controls, managements that focus on doing deals rather than managing, outright fraud and theft, and incentive systems that reward the wrong actions.

Enron followed this pattern. The case of Enron shows how a combination of intellectual laziness and groupthink by a large number of employees, consultants, and analysts allowed a group of greedy and ambitious individuals to get away with massive fraud. Enron was not a case of one or two people at the top undertaking a complex scheme unbeknown to others, but rather a case of many individuals who knew what they were supposed to do, but didn't do it. This was a case of analysts who never really questioned how Enron made its money, of accountants who didn't ask simple questions, and of employees and board members who saw dubious things but were afraid to stand up and ask the questions they should have.

Strategic Risk Management: A View

What is risk management, and how does it apply to the actions of employees? According to Kent D. Miller, "'risk' refers to variation in corporate outcomes or performance that cannot be forecast ex ante."[1] The key element here is to recognize that there is true uncertainty about human risk, or indeed any risk. The fact that an organization has survived to today without major scandal does not guarantee that it is safe in the future.

So what to do? According to Miller, effective risk management responses frequently include avoidance (which we have noted is almost impossible with the case of human risk), control (to be addressed in a moment), and cooperation and imitation (which can be achieved through quality initiatives).

Quality Initiatives Can Help

An organization is only as good as its parts—in this case the human parts. One fractured link in the chain means one vulnerable corporation. The quality aspect of management can be evoked to work hand in hand with problem prevention, but it is all too often overlooked.

Typically quality applies to (but is not limited to) reducing or eliminating defects in manufactured products. Beyond this, management also needs to invoke quality principles that smooth the internal environment. When intra-corporate communication channels are damaged, the ensuing misinformation may foster rogue behavior within the organization. Many quality experts cite training, transparency, empowerment, and clear communication as vital steps in building a quality organization.

Whether dealing with production issues or those relating to customer service, quality initiatives espoused by management thinkers like Armand V. Feigenbaum, J. M. Juran, Philip B. Crosby, and Frank Gryna can help a business. Firms that include quality as a core value, and reinforce this value through everyday practice, have experienced reductions down to zero of defects on production lines, lower worker turnover, higher levels of worker empowerment through training, more worker satisfaction, greater productivity, and a positive outlook on the company. Valuing people as the key drivers of both quality and performance is important to a firm and can go a long way toward identifying rogues and frustrating their efforts.

Quality starts with managers. Being an ethical role model is a key function of any leader. And the good news is that nothing special has to be done to become such a positive model. However, when leadership falters it can open the door to a rogue hit, doing as much damage to the corporation as a rogue wave can do to a ship at sea. You have to work at good leadership.

But the emphasis on quality alone is not enough. Control mechanisms, including both financial and performance audits, are important for preventing and uncovering potential problems. The really effective tools are punishment and brandishing the legal arsenal available to the company. Such measures reassure the public. A corporation just can't hunker down to avoid embarrassment. Swift and fair measures will fill the void of those strategic management initiatives that fail to catch rogue employees and will serve as a heavy reminder to others who may be about to embark on a negative course of action.

To many, the idea of punishment seems to be a return to management's dark past in the days of command and control. This is not the case. Taking corrective action, including negative reinforcements and punishments, is a legitimate function of managers, just as much as positive reinforcements are. Corrective actions can include firings, admonishments, wage deductions, and suspension without pay. People in authority are chary about digging in their heels to fight for what is ethically and obviously right for fear of being politically incorrect, or worse, manifestly insensitive. Many in decision-making positions prefer a course of inaction because they lack the gumption required to stay the course. If a manager has documented proof (paper or electronic) of wrongdoing by an employee, and particularly in a unionized environment, there is little that a union can do to "rescue" the employee from receiving the appropriate reprimand, short of the union condoning such rogue behavior.

Conclusion

A manager faces many risks—from industry-wide risks such as currency and interest rate risks, to department-specific risks such as accounting and treasury risks. Most of these risks can be quantified, though we are finding out that many of the numbers assigned to these risks are little more than educated guesses. Unfortunately the identification, measurement, and quantification of human risk are difficult and challenging. In spite of our best efforts, and in spite of pundits who spout an arsenal of "proof" to the contrary, reliable numbers cannot be assigned to human risk. Nor can risk be completely eliminated from an organization. But quality initiatives and control mechanisms can go a very long way to minimize exposure.

How Effective Strategic Risk Management Can Identify Rogues

Making It Happen

- Learn to live with the uncertainty of any risk, especially human risk.
- Place renewed emphasis on what is already being done, including audits (financial and performance), internal financial controls, and clear financial reporting.
- Vigilantly tweak and enforce the control mechanisms already in place. Think about expanding and/or adding controls.
- Revisit your own role as a highly visible manager. Are corporate controls short-sighted, or are they clearly structured so as to prevent deceit, fraud, and rogues from doing future damage?
- Identify high-risk areas in your firm—from inventory to treasury areas. Think about safety and security measures in addition to internal controls.

More Info

Books:

Crosby, Philip B. Completeness: *Quality for the 21st Century*. New York: Dutton, 1992.

Feigenbaum, Armand V. *Total Quality Control*. 4th ed. New York: McGraw-Hill, 2004.

Gryna, Frank, M. *Quality Planning & Analysis: From Product Development Through Use*. 4th ed. New York: McGraw-Hill, 2000.

Hill, Charles W. L., and Thomas McKaig. *Global Business Today*. 2nd Canadian ed. Whitby, ON: McGraw-Hill Ryerson, 2009.

Juran, J. M., and Frank M. Gryna (eds). *Juran's Quality Control Handbook*. 4th ed. New York: McGraw-Hill, 1988.

Mintzberg, Henry. *Managers Not MBAs: A Hard Look at the Soft Practice of Managing and Management Development*. San Francisco, CA: Berrett-Koehler Publishers, 2004.

Articles:

Becker, David M. "Testimony concerning new regulatory tools to control the activities of rogue individuals in the financial services industries." Given before the Subcommittee on Oversight and Investigations and the Subcommittee on Financial Institutions and Consumer Credit, US House of Representatives, March 6, 2001. Online at: www.sec.gov/news/testimony/ts042001.htm

Boak, Joshua. "Rogue trader rocks firm: Huge wheat futures loss stuns MFGlobal." *Chicago Tribune* (February 29, 2008).

Clark, Andrew. "From ethical champion to rogue interloper—BP's American nightmare: Accidents and allegations of market fixing destroy environmentalist image." *Guardian (London)* (November 16, 2006). Online at: tinyurl.com/273svw2

Gunther, Will. "In the crosshairs: Limiting the impact of workplace shootings." *Risk Management* 55 (November 2008). Online at: tinyurl.com/7drc3me

Johnston, David Cay. "Staff says IRS concealed improper audits and rogue agent." *New York Times* (May 1, 1998). Online at: tinyurl.com/aqf9tr

Malakian, Anthony. "Internal controls need to be tightened." *Bank Technology News* (April 2008). Online at: www.americanbanker.com/btn/21_4/-349038-1.html

Prince, C. J. "To catch a thief: Employee fraud hits growing businesses hardest. Here's what you can do to make sure there's not a thief among you." *Entrepreneur Magazine* (September 2007). Online at: tinyurl.com/77wpu2r

Report:
KPMG. "An approach to mitigating rogue trading risks." 2008.

Website:
CBC News coverage of the Conrad Black affair:
www.cbc.ca/news/background/black_conrad

Note

1. Miller, Kent D. "A framework for integrated risk management in international business." *Journal of International Business Studies* 23:2 (June 1992): 311–331. Online at: dx.doi.org/10.1057/palgrave.jibs.84902704

Countering Supply Chain Risk
by Vinod Lall
School of Business, Minnesota State University, Moorhead, USA

This Chapter Covers

- Business strategies such as outsourcing, lean manufacturing, and just-in-time lead to efficiency gains but at the same time expose the supply chain to higher risks.
- There are different sources of risk in a modern supply chain. Recognizing and appropriately managing these risks is necessary for a glitch-free functioning of the supply chain.
- Supply chain risk management strategies must be holistic and integrated with the whole supply chain environment.
- Firms must have dedicated budget line items for supply chain risk management activities.
- Failure mode effects analysis (FMEA) can be used to assess supply chain risks.

Introduction

In March 2000, a fire at a Philips semiconductor factory damaged some components used to make chips for mobile phones. Ericsson and Nokia—two of Philips' major customers—responded to the event in very different ways. Ericsson decided to let the delay take its own course, while supply chain managers at Nokia monitored the situation closely and developed contingency plans. By the time Philips discovered that the fire had contaminated a large area that would disrupt production for months, Nokia had already lined up alternative suppliers for the chips. Ericsson used Philips as a sole supplier and faced a severe shortage of chips, leading to delay in product launch and huge losses to its mobile phone division.

Today's global supply chains are complex and lean while efficiently delivering products and services to the marketplace. These supply chains involve a rigid set of transactions and decisions that span over longer distances and more time zones with very little slack built into them. As a result they are susceptible to several types of risk. These risks include operational risk due to demand variability, supply fluctuations and disruption risk due to natural disasters, terrorist attacks, pandemics, and breaches in data security. Such risks disrupt or slow the flow of material, information, and cash, and put billions of dollars at stake due to stock market capitalization, failed product launches, and the possibility of bankruptcies. In the above example, Ericsson lost 400 million euros after the Philips semiconductor plant caught fire; another example occurred when Apple lost many customer orders during a supply shortage of memory chips after an earthquake in Taiwan in 1999. Supply chain executives and managers must visualize and have a clear understanding of these risks along the entire supply chain, starting from the sourcing of raw materials to the delivery of the final product or service to the consumer. Once these risks are identified, they need to be scored on the likelihood of occurrence, and their impact must be quantified. Resources must then be used to mitigate or eliminate elements of high risk.

Types of Supply Chain Risk
Supply chain risks can be classified into different types depending on their origin. These include supply risk, demand risk, internal risk, and external environment risk.

Supply risk: These are the risks on the supply/inbound side of the supply chain. Supply risk may be defined as the possibility of disruptions of product availability from the supplier, or disruptions in the process of transportation from the supplier, to the customer. A supplier may be unavailable to complete an order for a number of reasons, including problems sourcing necessary raw materials, low process yield due to increased scrap, equipment failure, damaged facilities, or the need to ration its limited product among several customers. Transportation disruptions occur while products are in transit and add to the delivery lead time. They may be caused by delays in customs clearance at borders, or problems with the mode of transportation, such as the grounding of air traffic.

Demand risk: Demand risk is the downstream equivalent of supply risk and is present on the demand/outbound side of the supply chain. It may be due to an unexpected increase or decrease in customer demand that leads to a mismatch between the firm's forecast and actual demand. Increase in customer demand leads to depletion of safety stocks, resulting in stock-outs, back orders, and the need to expedite. A fall in customer demand leads to increased costs of holding inventory and, inevitably, price reductions. Other sources of demand risk are dependence on a single customer, customer solvency, and failure of the distribution logistics service provider.

Internal risk: This is the risk associated with events that are related to internal operations of the firm. Examples include fire or chemical spillage leading to plant closure, labor strikes, quality problems, and shortage of employees.

External environment risk: These risk elements are external to and uncontrollable from the firm's perspective. Examples include blockades of ports or depots, natural disasters such as earthquakes, hurricanes or cyclones, war, terrorist activity, and financial factors such as exchange rates and market pressures. These events disrupt the flow of material and may lead to plant shutdown, shortage of high-demand items, and price increases.

Strategies for Supply Chain Risk Management
Strategies for managing risk must be a part of supply chain management and must include processes to reduce supply chain risks that at the same time increase resilience and efficiency. Firms typically use basic strategies of risk-bearing, risk avoidance or risk mitigation, and risk transference to another party. The goal of risk-bearing is to reduce the potential damage caused by the materialization of a risk, and to be successful requires that early warning systems be installed along the supply chain. The main goal of risk avoidance is to reduce the probability of occurrence of a risk by being proactive, while under risk transfer the potential impact of risk is transferred to another organization such as an insurance company.

Mitigating Supply Chain Risks
A firm could use strategic and tactical plans under four basic approaches to mitigate the impact of supply chain risks. These approaches include supply management,

Countering Supply Chain Risk

demand management, product management, and information management. The task of managing supply chain risk is difficult as approaches that mitigate one risk element can end up exacerbating another. Also, actions taken by one partner in the supply chain can increase the risk for another partner.

Supply Management

Supply risks can be reduced by building a web of internal and external sources. Strategically, firms should focus their core competencies on new products and ideas and the engineering necessary to reduce time-to-market. They should continue to manufacture strategic, high-value, long-life products that have relatively low demand volatility while outsourcing non-strategic, low-value manufacturing and logistics services. It is important to be very selective in building a strong web of vendors and closely managing the vendor network. For each new product, the firm must capitalize on the varying expertise of its vendor network and use expected time-to-market, quality level and price to select a vendor from the network.

Tactical plans under supply management focus mostly on supplier selection and supplier order allocation. For this, firms should develop a profile of their supply bases to get a more complete picture of the supply side of the chain. This profile should include a wide range of supplier information including the total number of suppliers, the location and diversity of suppliers, and flexibility in the volume and variety of supplier capacities. Analysis of these data will help firms identify vulnerabilities in their supply chains so they can strategize, create contingency plans, conduct trade-off analysis of issues such as single sourcing, and, if needed, identify and line up backup sources.

Demand Management

Strategic plans under demand management focus on product pricing, while tactical plans are used to shift demand across time, across markets, and across products. One product pricing strategy is called the "price-postponement strategy," whereby the firm decides on the quantity of the order in the first period and then determines the price in the second period after observing updated information about demand. Shifting demand across time is known as "revenue management" or "yield management," whereby firms usually set higher prices during peak seasons to shift demand to off-peak seasons. One technique for shifting demand across markets is called "solo-rollover by market;" this involves selling new products in different markets with time delays, leading to non-overlapping selling seasons. To shift demand across products, firms use pricing and promotion techniques to entice customers to switch products or brands.

As with the supply side, firms must also develop a profile of the demand side to analyze the outbound side of the supply chain. Analysis of the demand side will identify dangers such as those associated with overreliance on a single distribution center to serve a large market, or the risks of having a highly concentrated customer base.

Supply Chain Reserves Management

Firms can deal with supply chain risks by holding reserves of inventory and capacity in the supply chain. Managers must decide carefully on the optimal location and size of these reserves as an undisciplined approach may lead to increased costs and hurt the bottom line.

Risk Management in an Uncertain World

Product Management
Firms can look at their internal networks and develop a profile of their products, processes, and services. Analysis of data in this profile can help to determine if there is a good mix of products and services and if there are risks in processes such as those used for fulfilling orders.

Information Management
Information technology tools can be used to understand and manage risk better by providing visibility into planned events and warnings for unplanned events in the entire supply chain. Firms must manufacture low-risk products first and use improved forecasts to produce the riskiest products very close to the selling season. This requires the use of reliable data and better forecasting methods. Key members in the supply chain must have easy and timely access to accurate information on such measures as inventory, demand, forecasts, production and shipment plans, work in process, process yields, capacities, backlogs, etc. This offers more opportunities to all parties to respond quickly to sudden changes in the supply chain and requires the implementation of information technology solutions that interface business data and processes end to end.

The collaborative planning, forecasting, and replenishment (CPFR) model is often used to induce collaboration and coordination through information sharing between supply chain partners such as retailers and manufacturers. Under CPFR, the manufacturer generates an initial demand forecast based on market intelligence on products, and the retailer creates its initial demand forecast based on customer response to pricing and promotion decisions. Both parties share their initial demand forecasts and reconcile the differences to obtain a common forecast. Once both parties agree on the common forecast, the manufacturer develops a production plan and the retailer develops a replenishment plan.

Making It Happen
It is critical to have an easy-to-use tool to identify and manage supply chain risk. FMEA is a well-documented and proven risk management tool that is used to evaluate the risk of failures in product and process designs. It can be used to evaluate supply chain risk using the following process steps:

- **Step 1.** Identify the categories of supply chain risk.
- **Step 2.** Identify potential risks in each category.
- **Step 3.** Use a rating scale of 1–5 to rate the opportunity, probability, and severity for each risk. The opportunity score for a risk is the frequency with which it occurs. One-time risk events receive an opportunity score of 1, while commonly occurring risk events are assigned an opportunity score of 5. The probability score is the score for the expected likelihood that a risk event will actually happen, so high probability scores are used when the probability of a risk event occurring is large. The severity score indicates the level of impact if the risk materializes. Low-risk events cause a minimum impact on the supply chain and receive a low severity score. Risk events that have a significant impact on the supply chain in terms of cost, time, and quality are assigned a high severity score.

Countering Supply Chain Risk

- **Step 4.** For each potential risk, calculate the risk priority number (RPN) as RPN = Opportunity × Probability × Severity.
- **Step 5.** Use Pareto analysis to analyze risks by RPN. Pareto analysis is a formal technique used where many possible courses of action are competing for the attention of the problem-solver. The problem-solver estimates the benefit delivered by each action and then selects the most effective action.
- **Step 6.** Develop action plans to mitigate risks with high RPN.
- **Step 7.** Use another cycle of FMEA to reassess the risks.

Conclusion

The pursuit of new markets for products and of new sources for components is making supply chains longer and more complex. With this expansion comes increased risk, which may result in disruptions to the supply chain. These disruptions may be unexpected and statistically rare, but they must be understood, identified, and managed.

More Info

Books:

Chopra, Sunil, and Peter Meindl. *Supply Chain Management: Strategy, Planning, and Operations*. 4th ed. Upper Saddle River, NJ: Prentice Hall, 2009.

Sheffi, Yossi. *The Resilient Enterprise: Overcoming Vulnerability for Competitive Advantage*. Cambridge, MA: MIT Press, 2007.

Websites:

Council of Supply Chain Management Professionals (CSCMP): cscmp.org

Supply Chain Council (SCC): www.supply-chain.org

Building Potential Catastrophe Management into a Strategic Risk Framework
by Duncan Martin
Boston Consulting Group (BCG), London, UK

> **This Chapter Covers**
>
> » Most organizations recognize the need for a strategic risk framework. Such a framework typically identifies and analyzes the key strategic risks faced by the organization, such as competitive, regulatory, technological, demographic, and environmental changes.
> » If adopted at the highest level, an effective framework will drive resource allocation and, consequently, the ability of the organization to achieve its goals.
> » Many organizations fail to integrate the potential impact of catastrophes into their framework. Despite investing considerable time and energy into a risk management framework, this failure can result in large, unexpected losses. For example, a business might foresee, and mitigate, the entry of a new competitor but be caught off guard by a major flood that causes equal disruption and loss in value.
> » To avoid being blindsided in this way, best-practice risk management builds catastrophe risk management into the same framework as strategic (and other) risks. With resources directed at those risks that pose the greatest threat, the full spectrum of risks is measured and managed consistently, thereby underpinning long-run organizational success.

Definitions

What is catastrophic risk? In brief, catastrophic risk is: "stuff happens." More precisely, it is the risk of extreme damage and loss of life from a natural or human cause. Some unexpected, perhaps unexpectable, event occurs. Half a world away from its source in southern China, SARS kills 38 people in Toronto; a nuclear reactor at Chernobyl is driven into a state its designers never even imagined even as its operators disable critical safety features, and it explodes; events in the Middle East cause Britons to blow themselves up on the London Underground.

Strategic risk is also stuff happening, but from a business point of view. An ailing computer manufacturer trounces established consumer electronics firms by producing the killer portable music device, and then follows up with a mobile phone that is both revolutionary and beautiful; tiny car firms constrained by postwar, small island scarcity eliminate waste by worshipping quality, end up reinventing the entire manufacturing process, and brutally upend incumbents; Wall Street's best and brightest simulate endless market disruption scenarios except the one that finally happens—no bids and no offers. Result: total paralysis.

Beyond strategic and catastrophe risk, financial and operational risk are equally necessary if less glamorous parts of a fully functional risk framework. Only through the consistent identification, measurement, and management of the full spectrum of risks can an organization be sure that it meets its objectives successfully.

Core Concepts

More formally, there are four core concepts in risk: frequency, severity, correlation, and uncertainty.

An event is frequent if it occurs often. Most catastrophes are, mercifully, infrequent. Historically, there is a severe earthquake (seven or greater on the Richter scale) about once every 25 years in California. Hence, the frequency of big earthquakes in California is 1/25, or about 4% each year.

An event is severe if it causes a lot of damage. For example, according to the US Geological Survey, between 1900 and 2005 China experienced 13 earthquakes that in total killed an estimated 800,000 people. The average severity was 61,000 deaths.

Most people's perception of risk focuses on events that are low-frequency and high-severity, such as severe earthquakes, aircraft crashes, and accidents at nuclear power plants. Strategic risk also focuses on such low-frequency/high-severity changes, such as disruptive technologies or new entrants. However, a fuller notion of risk includes two additional concepts: correlation and uncertainty.

Events are correlated if they tend to happen at the same time and place. For example, the flooding of New Orleans in 2005 was caused by a hurricane; the 1906 earthquake in San Francisco also caused an enormous fire.

Estimates of frequency, severity, and correlation are just that: estimates. They are usually based on past experience and, as investors know well, past performance offers no guarantee of what will happen in the future. Similarly, the probabilities, severities, and correlations of events in the future cannot be extrapolated with certainty from history: They are uncertain.

The rarer and more extreme the event, the greater the uncertainty. For example, according to the US National Oceanic and Atmospheric Administration, in the 105 years between 1900 and 2004 there were 25 severe (category four and five) hurricanes in the United States. At the end of 2004, you would have estimated the frequency of a severe hurricane at 25/105 or about 24% per year, but there were four severe hurricanes in 2005 alone. Recalculating the frequency at the end of 2005, you would end up with about 27% per year (29/106). That's a large difference, and would have a material impact on preparations.

Which estimate is correct? Neither, and both: Uncertainty prohibits "correctness." Uncertainty is the essence of risk, and coping with it is the essence of risk management.

Both catastrophic and strategic risk management are thus predicting and managing the consequences of rare, severe, and potentially correlated events under great uncertainty.

Building Potential Catastrophe Management

Think, Plan, Do

Integrating catastrophe risk into strategic risk management requires a common conceptual framework. Best-practice risk management is—always and everywhere—a three-step process: Think, plan, do (Figure 1).

Figure 1. Think, plan, do

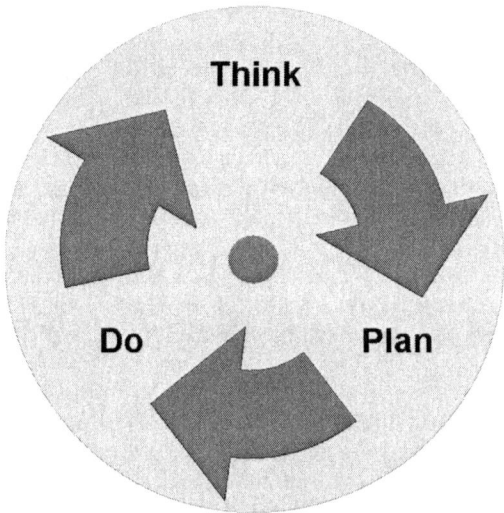

Think

Thinking comes first. Before being able to manage risk, risk managers must know how much is acceptable to their organization, and at what stage to cut any losses.

This risk appetite is not self-evident. It is a philosophical choice, an issue of comfort with the frequency, severity, and correlation of and uncertainty around potential events. Different individuals and organizations have different preferences.

Some people enjoy mountain climbing. They are comfortable with the knowledge that they're holding on to a small crack in a wet rock face with their fingertips and it's a long way down. Others prefer gardening, their feet firmly planted on the ground, their fingertips on their secateurs and not far from a cup of coffee. Similarly, some organizations aspire to blue chip, triple-A solidity, others to the rough and tumble of start-ups and venture capital, with the added drama of the San Andreas fault under their feet.

For strategic risk, managers attempt to simplify risk appetite down to how much money an organization is prepared to lose before it cuts its losses and changes objectives. For catastrophes, it is the frequency with which a certain event results in death—the frequency and severity of fatal terrorist attacks in London, for example. In some cases, it is defined externally. For example, on oil rigs in the North Sea it is defined through legislation. Events that cause death more often than once in 10,000 years are not tolerable and rig operators must mitigate the risk of any event with worse odds than this.

Risk Management in an Uncertain World

Plan

Planning is next. There are two parts: a strategic plan that matches resources and risks; and a tactical plan that assesses all the major risks identified and details the response to each one.

The first part is the big-picture risk appetite. If, for example, an organization decides that the frequency, severity, and uncertainty of flooding in London are too great, the big picture is that the organization needs to leave London, incurring whatever costs this requires. The strategic big picture also has to make sense. For example, although the high command of the US Army Rangers recognizes that they operate in very dangerous environments—occasionally catastrophically so, such as in Mogadishu, Somalia—and hence will on occasion lose soldiers, they have adopted a policy of "no man left behind." This helps to ensure that in combat Rangers are less likely to surrender or retreat, perhaps as a result winning the day. Consequently, governments spend a lot on flood defences, and armies spend a lot on search and rescue capabilities.

The next stage is detailed tactical planning. First, identify all the risks, strategic and catastrophic, financial and operational—all the things that might go wrong. Then, assess and compare them to see which are the most likely and the most damaging. Finally, figure out what to do, who's going to do it, and how much that's going to cost.

Many firms create business continuity plans on this basis. California's statewide disaster planning process is an excellent template for responding to catastrophes, because there's plenty of opportunity to practice: All manner of major incidents there—earthquakes, tsunamis, floods, wildfires, landslides, oil spills—occur relatively frequently. State law specifies the extent of mutual aid obligations between local communities and requires each to appoint a state-certified emergency manager. Emergency managers create a detailed disaster-management and recovery plan for their local community, reflecting local issues and needs. These plans are audited by state inspectors and rolled up into a statewide plan. To obtain the necessary resources, the plan is input to the state budgeting process.

Risk aversion does not necessarily make you safer. Many people or communities express a low risk appetite but baulk at the expense of reducing their risk to match their risk appetite. They don't put their money where their mouth is; instead they simply hope that the rare event doesn't happen. However, in the end, even rare events do occur. The results of mismatching risk appetite and resources were devastatingly demonstrated recently as Katrina drowned New Orleans.

Conversely, a large risk appetite is not the same thing as recklessness. Technology venture capital firms quite deliberately "bet the farm" on a few firms in narrow technology domains that they believe will be highly disruptive and hence profitable. This is high risk for sure, but the extensive deliberation and diligence of the investment and management processes mitigate the risk.

Do

Doing is a combination of activities. Before an event, *doing* means being prepared. This consists of acquiring and positioning the appropriate equipment, communications systems, and budget; recruiting, training, and rehearsing response teams; and ensuring that both the public and the response teams know what to do and what not to do. A contingency plan that is not tested is likely to fail.

After an event, *doing* means keeping your wits about you while implementing your plan, managing the inevitable unexpected events that crop up, and, to the extent possible, collecting data on the experience.

Once the epidemic has broken out or the earthquake has hit, the key is not to panic. Colin Sharples, a former acrobatic pilot and now the head of training and industry affairs at a British airline, observes that instinctively "your mind freezes for about ten seconds in an emergency. Then it reboots." Frozen individuals cannot help themselves or others. To counter this instinct, pilots are required go through a continuous and demanding training program in a flight simulator which "covers all known scenarios, with the more critical ones, for example engine fires, covered every six months. Pilots who do not pass the test have to retrain."

Most organizations operating in environments where catastrophes are possible have similar training programs, albeit usually without the fancy simulation hardware. In addition to providing direct experience of extreme conditions, such training also increases skill levels to the point where difficult activities become routine, even reflexive. Together, the experience and the training allow team members to create some "breathing space" with respect to the immediate danger. This breathing space ensures that team members can play their part and in addition preserve some spare mental capacity to cope with unexpected events.

The importance of this "breathing space" reflex reflects a truth about many extreme situations: They don't usually start out that way, but a "chain of misfortune" builds up where one bad thing builds on another and the situation turns from bad to critical to catastrophic. First, something bad happens. For example, first a patient reports with novel symptoms and doesn't respond to treatment. Then they die…then one of their caregivers dies too. Then one of their relatives ends up in hospital with the same symptoms…and so on. A team with "breathing space" can interrupt this chain by solving the problems at source as they arise, allowing them no time to compound. In this case, a suspicious (and perhaps even paranoid) infectious disease consultant (the best kind) might isolate the patient and implement strict patient/physician contact precautions before the infection was able to spread.

For most organizations, the critical learning point is not to create a continuity or contingency plan and then let it sit on a shelf. A plan that gathers dust is a dead plan—only living plans can save lives.

Close the Loop
When the *doing* is over and the situation has returned to normal, risk managers must close the loop and return to *thinking*. The group has to ask itself: "So, how did it go?" Using information collected centrally and participants' own experience, each part of the plan is evaluated against its original intention. This debrief can be formal or informal, depending on what works best. Sometimes it might even be public, such as

Risk Management in an Uncertain World

the Cullen inquiry into the disastrous Piper Alpha North Sea oil platform fire in 1989 that cost 165 lives.

Where performance was bad, the group must question whether the cause was local: training, procedures, and equipment; or strategic: the situation was riskier than the organization wants to tolerate, or is able to afford. These conclusions feed into the next round of *thinking* and *planning*.

Pitfalls

The main pitfall in the integration of catastrophe risk into strategic risk management is an insufficiently holistic process. Usually, this stems from the separation of strategy development, risk management, and, in many cases, insurance. In many organizations strategy development is the sexiest assignment and is jealously guarded by its departmental owners. As a result, sometimes strategic plans can be insufficiently informed by risk assessment. Since they tend to communicate in jargon and equations, risk management departments often do not help themselves. Sometimes, insurance is not a component of the risk management scheme; it is part of the finance area, and an obscure part at that. As a result, decisions on which risks to cover and to what degree may be taken without consideration of the organization's overall risk appetite. This lack of integration of the risk assessment process can ultimately lead to inconsistent treatment of risks and misallocation of scarce resources.

Case Study

Morgan Stanley was until recently a leading American investment bank. Investment banking is not for the fainthearted since it involves taking very large financial risks. Consequently, Morgan Stanley invested very large amounts in financial risk management. In general, this worked well and the firm was mostly profitable through the 1990s.

Managing financial risk is merely par for the course for investment banks. One thing that set Morgan Stanley apart from its peers was its assessment of catastrophe risk at one of its major operational hubs, the World Trade Center (WTC) in downtown New York. The corporate security manager, a decorated former soldier named Rick Rescorla, predicted the 1993 WTC bombing and had been able to convince the firm that such an attack would happen again; the firm had committed to move out at the end of its lease in 2006. On September 11, 2001, Morgan Stanley had 3,700 employees in the WTC. All but six—one of them Rescorla—got out alive, a direct result of constant practice and calm execution.

The integration of catastrophe risk into the strategic risk framework of the firm saved many lives. Few cases are this dramatic, but the point is the same: Risks are risks, regardless of source. The way we label them is entirely arbitrary. If, because of that labeling, we fail to treat all risks consistently, the consequences can be serious.

Building Potential Catastrophe Management

Making It Happen
In terms of implementation, there are five key principles.

- First, integration must be top down. Only senior management can both view the full holistic picture and require compliance further down.
- Second, the integration has to be genuinely "lived" by the senior managers. If employees feel that integration is merely lip service, they will not participate and the experiment will fail.
- Third, since risk appetites tend to be low with respect to very severe events, the resultant scarcity of events may drive hubris: Since it hasn't happened for a while, it probably won't or can't happen again. In industrial settings, researchers have observed that the odds of a serious accident increase with the time elapsed since the last one. Avoiding this complacency is critical.
- Fourth is the balance between sounding the alarm and having people respond. The more often an alarm sounds, the more likely it is that individuals will assume it's just a drill, or faulty, and tune it out, but if an alarm never sounds, no one will know what to do.
- Finally, many risk issues are amenable to sophisticated mathematical and computational treatments. There is a temptation to assume that just because a risk is measured, it is managed. It isn't.

More Info

Books:
Abraham, Thomas. *Twenty-First Century Plague: The Story of SARS*. Baltimore, MD: Johns Hopkins University Press, 2005.
Cullen, Lord W. Douglas. *The Public Inquiry into the Piper Alpha Disaster*. London: The Stationery Office, 1990.
Junger, Sebastian. *The Perfect Storm: A True Story of Men Against the Sea*. London: HarperCollins, 2007.
Perrow, Charles. *Normal Accidents: Living with High-Risk Technologies*. Princeton, NJ: Princeton University Press, 1999.
Pyne, Stephen. *Year of the Fires: The Story of the Great Fires of 1910*. London: Penguin, 2002.
Singer, P. W. *Corporate Warriors: The Rise of the Privatized Military Industry*. Ithaca, NY: Cornell University Press, 2004.

Article:
Stewart, James B. "The real heroes are dead." *New Yorker* (February 11, 2002). Online at: tinyurl.com/iggtkv

Websites:
California Emergency Management Agency (CalEMA): www.calema.ca.gov
Federal Emergency Management Agency (FEMA): www.fema.gov
London Prepared: www.londonprepared.gov.uk

Business Continuity Management: How to Prepare for the Worst
by Andrew Hiles
Kingswell International, Oxfordshire, UK, with international operations

This Chapter Covers

- No organization is immune from disaster.
- Business continuity management (BCM) is an integral part of corporate governance.
- A business continuity plan (BCP) can protect your brand, reputation, and market share.
- The prerequisite discipline of risk and impact assessment reveals critical dependencies and threats to them, enabling preventative measures to be taken.
- Risk and impact assessment identifies and prioritizes mission-critical activities and the timeframe in which they must be resumed; it can also provide new risk insights to improve your business performance.

Introduction
Over five years even a well-managed organization has an 80% chance of suffering an event that damages its profits by 20%.[1]

The cause could be equipment downtime, failure of utilities or supply chain, terrorism, fire, flood, explosion, or adverse weather. Whatever the cause, without a business continuity plan (BCP), the result is the same: damage to reputation, brand, competitive position, and market share. Sometimes this damage, and subsequent losses, are severe enough to lead to permanent closure.

Yet such loss can be minimized, or even avoided, by implementing a business continuity management (BCM) system which includes developing a BCP.

Quite simply, those organizations that have a BCP tend to survive a major adverse incident, while those without a BCP tend to fail.

What Is BCM?
According to one definition, BCM is: a "holistic management process that identifies potential impacts that threaten an organization and provides a framework for building resilience and the capability for an effective response which safeguards the interests of its key stakeholders, reputation, brand and value creating activities."[2]

Information and communications technology (ICT) disaster recovery is an important and integral part of BCM—but only one part. BCM covers all mission-critical activities (MCAs)—operations, manufacturing, sales, logistics, HR, finance, etc.—not just the technology.

There are several standards to provide guidance on BCM, the most recent—and probably the most universally accepted—is ISO 22301:2012, Societal Security—Business

Continuity Management Requirements, which is based on, and replaces, the well-established and respected BS 25999 BCM standard.

The BC Project

BCM starts as a project, but, once the BCP has been developed, audited, and exercised, it becomes an ongoing program needing regular maintenance and exercise.

The project activities are illustrated in Figure 1.

Figure 1. BCP project structure

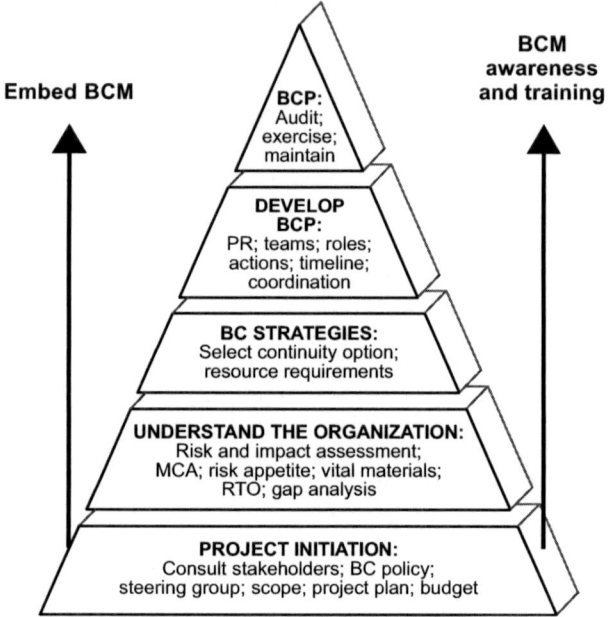

Making BC Happen
Phase One

The BC project should start with a clear understanding of the needs of the stakeholders and the support of the board. BC policy needs to be set.

A high-level steering group needs to be set up to decide priorities and define the scope of the project. For instance, is the objective to be "business as usual"—or will it just cover the 20% of goods or services that generates 80% of the profits? Will it cover all customers, or just the most important ones? Does it embrace all locations, or just head office? How far does it go down into the supply chain? Will it cover only local disasters, or is it to cope with wide-area disasters—hurricanes, major floods, etc.?

Next, a project plan should be developed, identifying the milestones and deliverables of the project. These include:

Business Continuity Management: How to Prepare for the Worst

- risk and impact assessment;
- agreeing BC strategies;
- developing the BCP and implementing contingency arrangements;
- audit and exercising the BCP.

A budget can be established for Phase One from a knowledge of how many sites are to be covered, how many people are to be interviewed, how many processes are to be included at each site, and an assessment of time for research and report-writing.

Risk and impact assessment can be broken down into subactivities:

- identification of assets and threats to them;
- weighting threats for probability and impact (in cash and non-cash terms);
- identification of MCAs and their dependencies;
- establishing the recovery time objective (RTO) for each (the maximum acceptable period of service outage);
- establishing the recovery point objective (RPO) for each (the timestamp to which data and transactions have to be recovered);
- identifying the resources needed for recovery and the timeframe in which they are required;
- identifying any gaps between the RTO, RPO, and actual capability (for example, the IT backup method may not permit recovery within the RTO);
- establishing the organization's appetite for risk;
- making recommendations for risk management and mitigation;
- making recommendations to close any gaps revealed.

The risk and impact assessment is usually conducted by analysis of building plans and operational layouts; review of reports on audit, health, safety, and environmental and operational incidents; interview of key personnel; and physical inspection.

Once these activities have been completed, possible contingency arrangements can be considered. The instinctive reaction is to replicate existing capability—but there may be more cost-effective options.

Holding buffer stock could cover equipment downtime. Increased resilience and "hardening" of facilities may reduce risk to an acceptable level. Items or services could be bought in, rather than undertaken in-house. Contracts could be placed with commercial BC service vendors for standby IT, telecommunications, and work area recovery requirements.

The risk and impact assessment then forms the basis for a cost–benefit analysis of the contingency options and allows a BC strategy to be recommended and agreed.

This report, incorporating the findings and recommendations from the risk and impact assessment, forms a natural closure to Phase One. Usually there is a natural break while recommendations are considered and the budget for Phase Two is agreed.

Risk Management in an Uncertain World

Phase Two

Once the BC strategy has been agreed, the BC plan can be started, bearing in mind what constraints may be placed on your organization by emergency services, public authorities, regulators, and landlords and other occupants (if you occupy a building with more than one tenant).

Incident and emergency management plans (for instance, evacuation, fire, bomb threat, etc.) need to be consistent with the BCP, and there needs to be escalation processes from them into the BCP. Triggers should also be identified for escalation from customer complaints, failure of service-level agreements, problem and incident management processes, etc., into the BC process.

The BC organization may not necessarily mirror the normal organization—for instance, multidiscipline teams may be appropriate—and the BC manager or coordinator may not usually hold the level of authority they are accorded under disaster invocation.

Typically the board will be separated in two: one to manage the ongoing business, the other to deal with the disaster situation. The emergency, crisis or business continuity management team (BCMT) will include board-level decision-makers. These include members from business and support units, and the BC manager (effectively the project manager for recovery) will report to them.

Business and support unit teams, including ICT, will report on recovery progress and seek clarification, information, and support from the BC manager. The BC manager will resolve any priority clashes within his or her authority and refer others to the BCMT.

Table 1 is a partial example of a BC organization. Additional BC teams will be created as necessary to cover each MCA, business, or support unit. The overview at Table 1 needs to be amplified by detailed action plans covering each BC team.

Table 1. Partial example of a BC organization

BC Management team	IT team	Base site recovery team
Leader: BC management team leader Alternate	*Leader:* TBD Alternate: TBD	*Leader:* TBD Alternate: TBD
Members: CFO Alternate COO Alternate PRO Alternate Marketing director Alternate Estates manager Alternate: TBD Admin support: TBD	*Members:* Applications manager Alternate PC servers/LAN manager Alternate Data/voice communications manager Alternate: TBD Admin support: TBD *Roles:* Recovery of all platforms, systems applications and data at standby site: TBD Data/voice communications recovery at standby site: TBD	*Members:* Office services manager Alternate PC servers/LAN Alternate: TBD Data/voice communications Alternate: TBD Damage assessment/salvage Alternate Loss adjuster: TBD Admin support: TBD

Business Continuity Management: How to Prepare for the Worst

BC Management team	IT team	Base site recovery team
Reports: BC manager Alternate *Roles:* Consider group (corporate) impacts Manage recovery Coordinate all team action Consider safety, security, and environmental issues Decide on priorities Reassure media and authorities		*Roles:* Damage assessment, limitation, and salvage Recovery at base site Recovery of operational capability at base site IT, data/voice communications recovery at base site

TBD: to be determined

The BCP coordinator is not necessarily the same person who will be BC manager once the BCP is completed. The BCP coordinator's role is to ensure that all BCPs are completed consistently and comprehensively.

The BCPs should not be scenario-based, since the disaster is unlikely to fit neatly into any scenario envisaged. Instead, they should be based on a worst-case scenario: total loss of MCAs. However, if they are developed in a modular fashion, only that part which is relevant need be invoked if a lesser disaster happens.

The BCP coordinator will draft a BCP for the BCMT and for his or her BC activities, including BCP invocation procedure, and will provide advice and guidance to the business and support unit BC coordinators.

Next, a template BCP should be developed that can be used for each team. Once they have had training, BCP development coordinators for each business and support unit complete these. A support program can be created for their guidance as they develop their BCPs.

Each BCP should spell out assumptions so they may be challenged (for example, an assumption that more than one site will not suffer a disaster at the same time; or that skilled people will be available post-disaster).

The minimum content should include:

- prioritized MCAs and a credible action plan for their recovery within RTO and RPO;
- lists of team members, alternates, roles, and contacts;
- resource requirements and when and how they are to be obtained;
- contact details of internal and external contacts;
- information on relevant contracts and insurance;
- reporting requirements;
- instructions on handling the media;
- any useful supporting information (such as damage assessment forms; maps and information about alternate sites; detailed technical recovery procedures).

Risk Management in an Uncertain World

Once the BCPs have been developed they can be audited, reviewing each BCP for comprehensiveness, clarity, and accuracy. This also ensures that interrelationships between BCPs are reflected in the counterparty BCP.

Rigorous exercises probe BCP effectiveness under different disaster scenarios and provide realistic training for BC team members.

Lessons from BC audit and tests should be incorporated into the BCPs. Where this has not yet been done, a list should be provided at the beginning of the BCP stating what weaknesses were found to exist; who is responsible for rectifying them; and the timeframe for doing so.

The BCP may take many forms: hard copy; handheld devices; memory sticks, etc. Whatever the format, it should be kept secure, and steps should be taken to ensure that only the current version can be held.

Case Study

Buncefield

Buncefield Oil Storage Terminal supplied fuel to London Heathrow from pipelines transporting fuel from the north of England. It was owned by Hertfordshire Oil Storage Ltd, a joint venture between Total and Texaco. Other businesses were attracted to the site—Marylands Industrial Park—because of its low cost.

Around 06:00 hours on Sunday, December 11, 2005, an explosion occurred, measuring 2.4 on the Richter scale; it was heard as far away as France and the Netherlands.

The Buncefield incident was the biggest explosion, and the accompanying fire was the biggest fire, in peacetime Europe. Twenty-five different fire services tackled the blaze with 600 fire fighters.

The explosion and subsequent fire:

- destroyed some 5% of UK petrol stocks and destroyed 20 fuel tanks;
- injured 200 people; 2,000 were evacuated;
- damaged more than 300 houses and required ten buildings to be demolished;
- caused all the schools in the county to be closed;
- cost local businesses and local authorities £1 billion: it impacted 600 businesses and prevented 25,000 staff from getting to work;
- disrupted global air traffic schedules and local transport;
- caused businesses to suffer disruption of supply;
- caused many organizations to invoke their BC plans;
- made big retailers reassess their supply chain issues;
- forced companies to make public statements to protect their share value;
- created major environmental impact from millions of gallons of burning oil, which required more than three million gallons of contaminated firewater with up to 40 different contaminants to be disposed of; it took 500 tankers five weeks to move it.

Business Continuity Management: How to Prepare for the Worst

Other impacts were equally devastating:

- By January 10, 2006, data recovery and communications restoration was still ongoing.
- By January 11, 2006, 75 businesses employing 5,000 people were still unable to use their premises.
- Insurance cover was inadequate to cover losses.
- In August 2006, 2,700 claimants sued for £1 billion in a case that will cost £61 million.
- Brewers Scottish & Newcastle lost £10 million of stock.
- Retailer Marks & Spencer closed a food depot, disrupting deliveries to retail outlets.
- Fujifilm, 3Com Corporation, and Alcom buildings were damaged.
- Andromeda Logistics' distribution centre was evacuated: operations resumed on December 12 from their alternative distribution center.
- Shares in British Petroleum, a bystander, briefly dived.
- ASOS (As Seen On Screen), an online fashion retailer, lost its new warehouse with £5.5 million stock (19,000 orders were refunded).
- British Airport Authority rationed aviation fuel at Heathrow: airlines diverted to other European airports to refuel.
- Broadcasts on BBC radio and television news urged motorists to avoid panic buying of fuel.
- The HQ of XL Video, a video producer for trade shows, events, television, and concerts suffered structural damage. They had 12 projects to load on the Monday morning. Their BCP diverted projects: all shows were shipped on December 12.
- IT outsourcing company Northgate Information Solutions Ltd had backups ready for collection at 07:00 hours daily, but the fire happened at 06:00. Local tax payments went uncollected, and billing information for utility companies was lost.

Hertfordshire County Council's crisis management plan worked: it had been used at the two rail incidents at Potters Bar and Hatfield and been thoroughly tested in October 2005.

Conclusion

Wise executives have long known the importance of risk and impact assessment and the need for contingency planning. With today's threats, this has never been more important. Buncefield proved the need to:

- develop a BCP to protect reputation, brand, and share value and market share;
- communicate to key stakeholders;
- keep investors and customers informed;
- have alternative sites for operations and for a control center;
- read and understand the emergency plans of the local authorities;
- ensure that key standby resources are in place, such as information (status, contacts); accommodation (operations and work area); and reserves (stock, spare equipment, etc.).

Buncefield cost local businesses £70 million, much of it uninsured. It is imperative to check insurance cover. The impact of a major disaster could last for months, or even years.

More Info

Books:

Hiles, Andrew. *Business Continuity: Best Practices—World-Class Continuity Management*. Brookfield, CT: Rothstein Associates, 2007.

Hiles, Andrew. *The Definitive Handbook of Business Continuity Management*. 3rd ed. Chichester, UK: Wiley, 2010.

Hiles, Andrew N. *Enterprise Risk Assessment and Business Impact Analysis: Best Practices*. Brookfield, CT: Rothstein Associates, 2002.

Von Roessing, Rolf. *Auditing Business Continuity—Global Best Practices*. Brookfield, CT: Rothstein Associates, 2012.

Websites:

Association of Contingency Planners (ACP): www.acp-international.com

Business Continuity Institute (BCI): www.thebci.org

Continuity Central: www.continuitycentral.com

Disaster Recovery Institute (DRI) International: www.drii.org

Standards:

BS 25999 Business Continuity Management (UK)

HB 221 Business Continuity Management (Australia)

NFPA 1600 Emergency Management and Business Continuity (US)

ISO 22301 Business Continuity Standard

Glossary

BC:	Business continuity
BCM:	Business continuity management
BCP:	Business continuity plan
BIA:	Business impact assessment
DRP:	A plan for the continuity or recovery of information and communications technology (ICT)
MCA:	Mission-critical activities
Risk appetite:	The level of loss that an organization is prepared to tolerate
RTO:	Recovery time objective
RPO:	Recovery point objective (of data or transactions)

Notes

1. Oxford Metrica, www.oxfordmetrica.com
2. British Standards Institute/Business Continuity Institute Publicly Available Specification 56.

Risk Management: Beyond Compliance
by Bill Sharon
SORMS, USA

This Chapter Covers

- The boundaries between risk management and compliance have eroded over the past decade, to the detriment of both functions.
- The definition of risk should be expanded to include opportunities and uncertainties, not just hazards.
- The context for assessing operational risk is business strategy.
- The role of risk managers needs to expand so that they become coordinators of the risk information that is readily available in operational and business units.
- The perception of risk is dependent on one's organizational responsibilities, and the convergence of those perceptions is the central focus of the management of risk.

Introduction

Over the past decade the line between risk management and compliance has been blurred to the point where, in many organizations, it is impossible to determine if they are not one and the same. In part, this confusion between the two functions was initiated and then exacerbated by the passage of the Sarbanes–Oxley Act of 2002 and the implementation of Basel II. Both of these events consumed a great deal of resources, and many consulting firms labeled these efforts "risk management." They are, in fact, compliance requirements designed to protect stakeholders and, in the latter case, ensure the viability of the financial system. They are not designed for, and nor can their implementation achieve, the management of risk in individual companies or financial institutions.

This confusion between compliance and risk management has led to a defensive posture in dealing with the uncertainties of the competitive business environment. Risk has been confined to the analysis of what could go wrong rather than what needs to go right. Risk management organizations have become the arbiters of what constitutes risk and have assumed an adversarial relationship with business managers, particularly in capital allocation exercises. Failures and scandals are met with calls for more regulation, the implementation of regulations becomes the province of risk management organizations, and the execution of strategy (arguably the area in most need of risk management) becomes further separated from any kind of disciplined analysis.

An Expanded Definition of Risk

As Peter Bernstein tells us in his book *Against the Gods: The Remarkable Story of Risk*, the word risk comes from the old Italian *risicare*, which means "to dare." Daring is the driving idea behind business, the idea that a product or a service can achieve excellence and value in the marketplace. Strategy necessarily incorporates risk from the perspective of those actions which are required for its success.

Risk Management in an Uncertain World

In 1996 Robert G. Eccles, a former Harvard Business School professor, and Lee Puschaver, a partner at Price Waterhouse (now PricewaterhouseCoopers), developed the concept of the "business risk continuum." They argued that organizations that were successful in managing risk were those that focused on uncertainties and opportunities as much as they did on hazards. The context for evaluating risk in this manner is business strategy. This idea—that the definition of risk should be expanded to include those actions that an organization needed to embrace to achieve its goals—was revolutionary and codified what some companies were already beginning to initiate. Unfortunately, the narrow view of risk has prevailed for the past decade, and Eccles and Puschaver's work has essentially been ignored.

The overwhelming emphasis of most risk organizations today is on the hazard end of the scale. Dot.com, Enron, and now subprime, along with the increased focus on terrorism, cataclysmic natural disasters, and the potential for pandemic diseases, have placed most complex organizations in a defensive posture. The problem with this approach is that risk driven from the hazard perspective is experienced as overhead in the operational disciplines and business units; it's a cost of business, not an activity that enhances value or improves the possibility of success.

By expanding the definition of risk (or returning to its original meaning) companies can harness the inherent risk management abilities and information available throughout their organization and develop a predictive process to address mission-critical tasks. Understanding how risk is perceived and how people react to those perceptions is an essential step in managing the opportunities and uncertainties inherent in implementing a business strategy.

Organizational Roles and the Perception of Risk

Daniel Kahneman and Amos Tversky, the authors of "Prospect Theory," conducted a variety of experiments on the perception of risk and the responses that people had to identical information presented in different contexts. Among their conclusions they determined that:

- emotion always overrides logic in the decision-making process;
- people suffer from cognitive dysfunction in making decisions because they never have enough information;
- people are not risk-averse, they are loss-averse.

While these conclusions may be unsettling to those involved in quantitative risk analysis, all three are useful assumptions around which to build a proactive risk management process. Emotion is at the core of any business—the desire to produce the best product, offer the best service, and compete in the marketplace comes from passion, not analytics. Managing risk is about managing emotion, not eliminating it.

From an organizational perspective, the perception of risk is colored by one's responsibilities. In the operational environment, technologists see opportunities in deploying software and hardware. HR professionals define success as the attraction and retention of high-performance employees. In business units, opportunities require risks to be taken in order to capture market share or evolve a product line to the next level. Often these business

Risk Management: Beyond Compliance

leaders are unaware of the operational capabilities and capacities on which they must rely to achieve their goals. Operational managers often lack clarity on the business models they support. Individually, these perceptions of risk tell only part of the story and require the balance of all of the organizational perceptions in order for the cognitive dissonance to be managed and mitigated.

In this context, risk managers become coordinators of business intelligence rather than arbiters of what is and is not a risk. The management of risk is a communication process that is central to the success of the enterprise rather than an overhead process that compliance so often becomes. Participation in risk management is equivalent to participating in the development of business strategy. The desire not to lose (rather than the misguided view of being averse to "daring") is the underlying motivation for the process.

The Risk Perception Continuum

Figure 1. The risk perception continuum

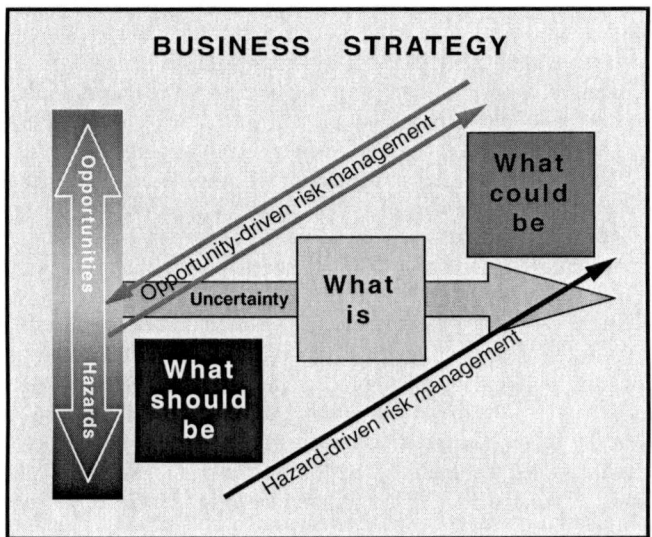

The risk perception continuum (Figure 1) summarizes the categories of risk and how they can be placed in an operational context. Using Eccles and Puschaver's concept of the three categories of risk, an organization can assign one of three different perceptions to determine the source and value of risk information:

>> **What Should Be** is the perception of risk that comes from external standards. These are "best practices" for both operational and business managers. The measures involved determine the degree to which an organization is aligned with these practices in the context of what the organization wants to achieve. For example, alignment with "best practices" for a data center is likely to be more important for a financial institution than an advertising agency. It is tempting to place compliance functions

47

in this area and track these issues as hazards. This is a mistake on two levels. First, the risk management process is central to the success of the organization and needs the oversight of the audit function. Putting them in the same unit creates a conflict of interest, one that is clearly identified in the Committee of Sponsoring Organization's (COSO) enterprise risk management framework. Second, compliance is a legal and regulatory function. One does not assess the risk of not complying. The primary audiences for this information are regulators and external auditors, and the ability to adhere to these requirements is really the baseline for participating in the marketplace.

▸ **What Is** comprises the uncertainty of the operating environment of the organization. This is the area where quantitative analysis and hedging are done to determine the upside and downside of a deal. It is here that both business and operational managers have the greatest impact on the management of risk, and it is here that the communication of the different perceptions of risk is most critical. The convergence of these perceptions constitutes valuable business intelligence. The classic example of managing risk in this manner is the HR hiring process. The MD of equity trading in an investment bank may have an urgent need for a large number of junior traders. The human resources department has a responsibility to ensure that the people the MD wants to hire have actually attended the universities claimed on their resumés and that they have passed a strenuous background check. The tension between these two perceptions is satisfied by the candidates signing a letter accepting their immediate dismissal should they be found to have misrepresented their qualifications. The organization embraces the risk that the contributions to the strategy will outweigh the potential for any damage that might be done during a relatively small window of time.

▸ **What Could Be** is the repository of the strategy of the organization and the perception of what risks need to be taken for it to be achieved. This perception is dynamic and responds to the demands of the marketplace, as well as the capabilities of the operating environment. Perhaps the best known example of how strategy drives the management of risk in an organization is the behavior of the US space agency, NASA, following John F. Kennedy's announcement that there would be an American on the moon by the end of the 1960s. In recently released tapes of meetings between Kennedy and James Webb, the director of NASA, the impact of strategy on operational capabilities is well illustrated. Webb advises Kennedy of the vagaries of space and the need to expand the space program to include a number of interim steps necessary to gain a better understanding before anyone can go to the moon. Kennedy listens and then tells Webb that he doesn't care about space, he wants to get to the moon before the Russians. What's interesting about this exchange is that Kennedy was defining a strategic goal that had no near-term likelihood of being achieved. He was also using that strategic goal to redefine the risk. The technical risk was unknowable at the time, but the political risk was quantifiable. Strategy organizes the operational environment and focuses it in specific directions. It requires operational managers to converge their perceptions of risk with the goals of the organization.

Risk Management: Beyond Compliance

Figure 2. Converging the perceptions of risk

Figure 1 also demonstrates the difference between driving risk management from the opportunity or strategy perspective as opposed to the hazard perspective. The latter approach tries to force standards up through the organization. Operational managers experience this as an audit process and, other than quarterly reports from the audit committee, very little of this information receives much attention from the senior executives responsible for implementing strategy.

Alternatively, risk management driven from the opportunity perspective creates a communications vehicle for the entire organization. This is a bi-directional process because, as the strategy is communicated into the operating environment, the organization responds with business intelligence.

Implementing a Risk Management Process

Using the organization's strategy as the context (rather than "best practices" or regulatory requirements), the first step in the process is to ask operational managers to identify the risks that must be embraced in order to achieve this strategy (operational disciplines are defined as those organizational units that do not generate income, i.e. finance, HR, IT, PR, etc.). Once identified, these activities are assessed—usually using a RAG (red, amber, green) rating—to determine the likelihood of their being achieved.

There are two important steps in this first stage of the process that are often lacking in risk management programs.

1. Operational managers are asked to predict a risk rating, usually on a quarterly basis, for the next four quarters. This provides the organization with more valuable data than point-in-time risk assessments, whose shelf-life

Risk Management in an Uncertain World

tends to be quite short. It also provides operational managers with the ability to communicate anticipated challenges in the future and/or illustrate how current challenges will be positively addressed over time.

2. Operational managers are also asked to note whether the activities they believe must be undertaken have sufficient funding. Once this information has been collated, the organization has a map of where it is investing in managing risks central to the strategy and where it is not.

Operational managers are then asked to complete an actual vs planned assessment at the end of each quarter. This is not an exercise to assess competency, but rather another channel for communication in the risk management process. Strategy may change, requiring a new perception of risk. Operational awareness of greater or lesser challenges may impact the original risk rating. Departures from the original assessment are expected and should be viewed as business intelligence rather than as a scoring of prescient abilities.

Once the process is established with the operational managers, the second stage of the risk management process can be implemented. Here, business managers are asked to contribute their perceptions of risk to the mission-critical operational activities that have been identified. For example, if the IT department identified the rollout of a new operating system as a risk that needed to be embraced and rated it as an amber or a red, given the exposure in maintenance and security, the business managers might rate it as a green as they have no clear knowledge of the technical issues. Differences in the perception of risk are expected and provide an opportunity to understand risk across operational and business disciplines.

The third stage (Figure 2) in the risk management process is the audit review, which not only validates the process itself, but also uses the risk assessments as a source for audit oversight of specific operational activities. The convergence of perception between operational and business managers and the audit function provides the risk management process with the widest possible range of understanding of risks to the strategy.

Once this process is established, metrics can be applied to risk ratings, operational disciplines can be weighted in importance by business unit, and portfolio views of risk can be developed across business units.

Case Study

JP Morgan—Managing the Risk of Outsourcing

The risk management process can be scaled to encompass the entire organization, a specific business unit, or a large project. A year prior to outsourcing 40% of its technology, JP Morgan initiated a predictive risk management program that converged the perceptions of technology and business managers and established an IT risk profile for each business unit.

▶▶ The IT self-assessment process was conducted quarterly on a global basis, and provided the bank with a portfolio view of IT operational risk across all business units.

Risk Management: Beyond Compliance

- The risk profiles allowed the bank to negotiate service levels based on an understanding of where the internal IT group was supporting the business strategy and where improvements were necessary.
- The IT self-assessment process was transferred to the successful vendors and the business units continued to contribute their perceptions, resulting in a shared process between the vendors and the bank.
- Perhaps the most important result of the process was a better understanding in the business units of IT capabilities and capacities. The organization gained an understanding of the technology that provided a competitive advantage (and should therefore be retained in the bank) and of the infrastructure and shared applications that could be turned over to external vendors.

Conclusion

No risk management function can ensure that negative events won't happen. The complexity of the markets and the speed of change create exposures that are difficult to predict. Managing risk as a process that engages the entire enterprise in the achievement of the business strategy does, however, create a resilient organization that can better respond to difficulties that always arise.

Making It Happen

The operational risk management process described in this article begins with the business strategy but ultimately engages the entire organization. Senior management needs not only to endorse the process but also to participate in and use it on a continuing basis. The early stages of the process require patience, and some care should be taken in the initial implementation.

- There is often confusion in the operational disciplines about what is a risk to the business strategy and what is a best-practice or compliance requirement. Risk managers will likely need to assist operational managers in this distinction.
- Simplicity is key in the early stages of the risk management process. Many efforts collapse under their own weight when organizations attempt to accomplish too much in a short period. Risk management is about leveraging existing expertise; complex metrics can be applied once the system is robust.
- Using the risk management process as a communication process, not only for challenges but also for capacities and creative solutions, is essential in making it a robust vehicle for the generation of business intelligence.

More Info

Book:
Bernstein, Peter L. *Against the Gods, The Remarkable Story of Risk*. New York: Wiley, 1996.

Article:
Kloman, Felix. "Risk management and Monty Python, part 2." *Risk Management Reports* 32:12 (December 2005). Online at: tinyurl.com/34hwowx

Risk Management in an Uncertain World

Report:
Puschaver, Lee, and Robert G. Eccles. "In pursuit of the upside: The new opportunity in risk management." Leading Thinking on Issues of Risk, PricewaterhouseCoopers, 1998.

Websites:
COSO (Committee of Sponsoring Organizations of the Treadway Commission): www.coso.org

Prospect theory: prospect-theory.behaviouralfinance.net and www.sjsu.edu/faculty/watkins/prospect.htm

Risk Metrics: www.riskmetrics.com

Strategic Operational Risk Management Solutions (SORMS): www.sorms.com

How to Better Manage Your Financial Supply Chain

by Juergen Bernd Weiss
SEPA-Now Consulting, Frankfurt, Germany

This Chapter Covers

- Financial supply chain management (FSCM) addresses a number of initiatives that can help to make finance organizations more efficient and improve the working capital position of an enterprise.
- There are a number of indicators for an inefficient financial supply chain including low straight-through processing rates and a high amount of uncollectible receivables on the balance sheet.
- Key performance indicators such as days sales outstanding or days in receivables can be used by companies to benchmark themselves with their peers.
- Microsoft decided to improve its financial supply chain to better utilize working capital, to reduce bank fees, to process payments more effectively, and to gain better control of cash flows.

Introduction

Benchmarks of business performance indicate that enterprise resource planning (ERP) systems and other enterprise technologies have transformed customer and supply chain processes but that the performance of the finance function has hardly changed. Although some companies have managed to improve the performance of their financial processes profoundly, financial functions are still neglected in many businesses, and days sales outstanding (DSO) and working capital needs are very high in several industries. The working capital scorecard for 2011 from *CFO Magazine* demonstrates that there are significant differences between high and low performers within an industry. In the pharmaceuticals industry, for example, the best score in DSO was 48, while the worst score was 117—two times more than the sector median of 57. Research from the Hackett Group indicates that finance department costs continue to consume more than 1% of revenues in many companies, and CFOs struggle with poor transparency of their daily cash flows.

In times when unprecedented economic uncertainty and soaring stockholder expectations are putting every function under closer scrutiny than ever before, the finance function should be driving business, not holding it back. Financial supply chain management (FSCM) can help companies to remove some of the inefficiencies in operational processes in order to become more effective.

Definitions of Financial Supply Chain Management

There are different definitions of the term *financial supply chain*, which appeared for the first time in 2000 and 2001. According to the research company Killen & Associates (2001), the financial supply chain "parallels the physical or materials supply chain and represents all transaction activities related to the flow of cash from the customer's initial order through reconciliation and payment to the seller." The Aberdeen Group, another research company, calls the financial supply chain "a range of B-to-B trade-related intra- and inter-company financial transaction-based functions and processes [which] begin before buyers and suppliers establish contact and proceed beyond the settlement process." The two definitions emphasize different topics. Killen's focuses on the parallelism between the physical and the financial supply chain, and it stresses a section of the cash flow cycle that I'll discuss in more detail below. The Aberdeen Group's definition focuses on the collaborative nature of financial supply chain management and reveals that the financial value chain isn't limited to the inner walls of a company but includes communication and cooperation with business partners.

Both definitions focus on a process-oriented view of the financial supply chain that is basically correct; however, in many respects the explanations do not go far enough:

- They focus very much on the collaboration between companies—specifically, suppliers and customers—and they do not consider other important business partners within the financial supply chain, such as banks.
- They describe primarily the status quo, and do not stress the various dimensions for the optimization of business processes within the financial supply chain.
- The motivation, as well as the key performance indicators, for an efficient financial supply chain are not obvious.

Another definition that includes these three aspects is the following: Financial supply chain management (FSCM) is the holistic and comprehensive planning and controlling of all financial processes which are relevant within a company and for communication with other enterprises. The goal of FSCM is to increase the transparency and the level of automation of business processes along the financial value chain. The purpose is to save processing costs and reduce the working capital of the company. This definition doesn't consider where the financial supply chain actually begins and ends, because there are also analytical processes that are not directly related to a business process but which belong nonetheless to the financial supply chain. Let's now have a closer look at the indicators of an *inefficient* financial supply chain.

Indicators of an Inefficient Financial Supply Chain

As we have seen, the financial supply chain is different from the physical supply chain because it deals with the flow of cash instead of goods. Just as in the physical supply chain, though, every day that's lost in the cash-to-cash cycle equals lost revenue. But how do you know that your financial value chain isn't working properly? Besides a number of rather operational problems, there are also several concrete key performance indicators and metrics that you can use to analyze your financial supply chain. You are most likely aware of the fact that the financial supply chain stretches across many different business processes. These are, in a broader sense, the two processes *order-to-cash* and *purchase-to-pay*, which consist of various sub-processes that are relevant to the financial aspects of the value chain.

How to Better Manage Your Financial Supply Chain

The order-to-cash process includes, from the perspective of a supplier (or creditor), the following business process steps:

1. Creditworthiness check.
2. Invoice creation.
3. Cash forecast.
4. Financing of working capital.
5. Processing of dispute cases.
6. Cash collection.
7. Settlement and payment.
8. Account reconciliation.

From the perspective of a customer (or debtor), the purchase-to-pay process consists of the following business processes:

1. Procurement.
2. Cash forecast.
3. Financing of working capital.
4. Receipt of invoices.
5. Resolution of discrepancies or exceptions.
6. Invoice approval.
7. Settlement and payment.
8. Account reconciliation.

There are a number of operational factors within the order-to-cash and purchase-to-pay processes that can serve as indicators of a suboptimal financial supply chain. Some examples are:

- The number of paper-based business processes is very high and there are several changes in medium (for example, the creation of invoices).
- The straight-through processing rate is low, which means that there are multiple manual interventions and process steps.
- Companies struggle with a large number of dispute cases during the creation of invoices, and it takes them a lot of time to process these.
- There is a large amount of uncollectable receivables on the balance sheet, and many employees in receivables or collections management are involved in the resolution process.
- Enterprises haven't implemented a consistent credit management policy, which results in a number of bad debt losses.
- Management has difficulties in predicting cash flows.
- There is no centralized cash management to control payment streams, and the company maintains too many bank connections.

Key Performance Indicators

There are various key performance indicators that are relevant for measurement in financial supply chain management. One key metric is the cash flow cycle, which defines the period from delivery by suppliers until the cash collection of receivables from customers (Figure 1). It is the time period required for the company to receive the invested funds back in the form of cash. The cash flow cycle can be divided into the *operating cycle*—which is the time period between delivery by suppliers and the actual cash collection of receivables, and the *cash flow cycle*—which is the time period between the cash payment for inventory and the cash collection of receivables. The longer the cash flow cycle, the greater is the working capital requirement of a company, which means that a reduction of the cash flow cycle will immediately free up liquidity.

Figure 1. The cash flow cycle

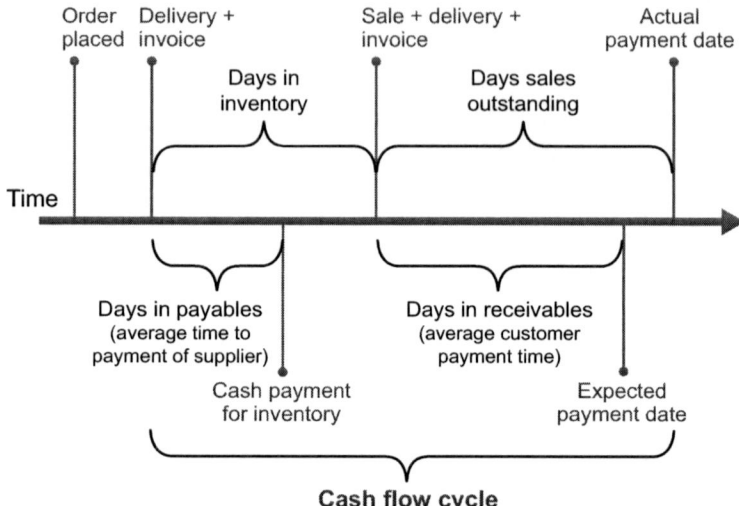

Within the cash flow cycle we can differentiate the following parameters, which are delimited in Figure 1 with curly brackets:

- ▸ Days in inventory: This is the length of time between the delivery of the goods and the invoice from the supplier, and the sale of the goods and the invoice to the customer. It describes the average number of days the goods of a company remain in inventory before being sold. This metric is the focus for all activities around classical supply chain management.
- ▸ Days in payables: This is the length of time between delivery of the goods and the invoice from the supplier, and the actual payment for the inventory. This figure describes the average time it takes to pay a supplier. The parameter considers the outstanding receivables of a company, and is an important metric for debtors concentrating on their efforts to optimize the purchase-to-pay cycle.
- ▸ Days sales outstanding: This is the length of time between the sale of the goods and the invoice to the customer, and the actual payment date of the customer. This metric measures the average number of days companies need to collect revenue after a sale has been made. A high DSO number means that an enterprise is selling to its customers on credit and taking longer to collect money. The figure is an important figure for creditors, to optimize the order-to-cash cycle.

How to Better Manage Your Financial Supply Chain

▶▶ Days in receivables: This is the length of time between the sale of the goods and the invoice to the customer, and the expected payment date. This key performance indicator is similar to DSO, and indicates the average time, in days, that receivables are outstanding. Days in receivables can also be called best possible DSO, since the company would collect all receivables before the due date.

Within the cash flow cycle there is potential to reduce both days in inventory and days sales outstanding. Days in payables can be reduced but should be monitored carefully to avoid putting supplies at risk. Days in receivables can be reduced by optimizing cash collection. Another important indicator for an efficient financial supply chain management is working capital, which is a balance sheet metric and part of the liquid assets. Working capital is calculated as current assets less current liabilities, and is a measure of the liquid reserve and short-term solvency of an enterprise, available to satisfy contingencies and uncertainties. One of the key objectives of financial supply chain management is to optimize the working capital by reducing, for instance, outstanding receivables.

Case Study

Microsoft

US company Microsoft decided to improve its financial supply chain by replacing third-party and in-house developed legacy software systems that were very costly to maintain. Microsoft, which is headquartered in Redmond, Washington, provides information technology, operating systems, middleware solutions, small/mid-size business applications, professional services, and other software solutions. The company reported annual sales in 2008 of more than US$60 billion and had more than 90,000 employees worldwide.
The main business drivers for the company were:

▶▶ Better data integration between applications.
▶▶ Elimination of manual intercompany processes and month-end bank account reconciliation.
▶▶ More transparent accessibility to real-time data such as bank account balances, financial transactions, and accounts receivable and payable.
▶▶ More efficient usage of excess funds and better working capital management.
▶▶ Increased straight-through processing of foreign exchange trading.
▶▶ Reduction of bank fees and more cost-effective processing of payments.
▶▶ Risk reduction and better control of cash flows.
▶▶ Better utilization of human resources.

Microsoft decided to implement a vendor's financial supply chain management solution to complement its existing enterprise resource planning (ERP) landscape. The company realized a number of benefits from the project and was, for example, able to automate the confirmation process in the foreign exchange settlement fully. Exception rates are now smaller than 5%, the settlement process went from four hours to less than 15 minutes, and the percentage of settlement errors is approaching 0%.

More Info

Books:

Bhalla, V. K. *Working Capital Management: Text and Cases*. New Delhi: Anmol Publications, 2006.

Horcher, Karen A. *Essentials of Managing Treasury*. Hoboken, NJ: Wiley, 2006.

Salek, John G. *Accounts Receivable Management Best Practices*. Hoboken, NJ: Wiley, 2005.

Schaeffer, Mary S. *Essentials of Credit, Collections, and Accounts Receivable*. Hoboken, NJ: Wiley, 2002.

Sagner, James. *Essentials of Working Capital Management*. Wiley, 2010.

Nash, Thomas. *Financial Supply Chain Management: The Next Wave*. Gower, 2012.

Articles:

Hartley-Urquhart, Roland. "Managing the financial supply chain." *Supply Chain Management Review* (2006). Online at: www.scmr.com/article/CA6376439.html

Katz, David M. "Easing the squeeze: The 2011 working capital scorecard." *CFO Europe Magazine* (July/August 2011). Online at: www.cfo.com/article.cfm/14586631

Websites:

CFO—News and insight for financial executives: www.cfo.com

gtnews—Library for finance and treasury professionals: www.gtnews.com

The Hackett Group: www.thehackettgroup.com

Managing Interest Rate Risk
by Will Spinney
Association of Corporate Treasurers, UK

This Chapter Covers

- Interest rate risk can manifest itself in several different ways.
- It is best managed within the context of the firm and a risk framework.
- Proper evaluation or measurement is key.
- Selection of a good key performance indicator is essential.
- A typical response to interest rate risk is a transfer of risk to another party.
- Many risk transfer tools are available, of which interest rate swaps are the most popular.
- The risk is usually transformed rather than eliminated.

Introduction
Almost all firms are exposed to interest rate risk, but it can manifest itself in different ways. A proper response to this risk can only come following a full understanding of the context of the firm and its strategy, along with a full evaluation of the risk. Firms should generate a well thought out key performance indicator (KPI) and then apply one or more of the many tools available in the market to transfer interest rate risk.

Major Ways That a Firm Can Be Affected
Interest rate risk is the exposure of the firm to changing interest rates. It has four main dimensions:

Changing Cost of Interest Expense or Income
Companies with debt charged at variable rates (for example, based on Libor, and also called floating rates) will be exposed to increases in interest rates, whereas companies whose borrowing costs are totally or partly fixed will be exposed to falls in interest rates. The reverse is obviously true for companies with cash term deposits. This is usually the key risk that firms consider.

Impact on Business Performance by a Changing Business Environment
Changes in interest rates also affect businesses indirectly, through their effect on the overall business environment. In normal times, for example, construction firms enjoy a rise in business activity when interest rates fall, as investors build more when the cost of projects is lower. Conversely, some firms may benefit from high levels of activity that prompt a high interest rate response by central banks. So some firms may have a form of natural hedge against the other forms of interest rate risk, although for any one firm the effect may lead or lag actual changes in rates.

Impact on Pension Schemes Sponsored by the Firm
Pension schemes that carry liability and investment risk for the sponsor have interest rate risk in that liabilities act in a similar way to bonds, rising in value as interest rates fall and vice versa.

Changing Market Values of Any Debt Outstanding
Although a nonfinancial firm will usually report its bonds on issue in financial statements at substantially their face value, early redemptions must be done at the market value. This may be significantly different, as interest rates will change the value of fixed-rate debt. This risk is not commonly considered by most nonfinancial firms.

Interest Rate Risk in the Context of the Firm
Investors do expect firms to take risks, especially with regard to their core business competencies. It may be that investors expect the firm to take interest rate risk. On the other hand, investors would probably not expect a firm to breach a financial covenant because of rising interest rates.

Risk Management Framework
A risk management framework includes the following key stages:

- Identification and assessment of risks;
- Detailed evaluation of the highest risks;
- Creation of a response to each risk;
- Reporting and feedback on risks.

Evaluation is crucial to the management of interest rate risk and will discover exactly how a firm might be affected, thus guiding the response to the risk. Evaluation techniques include: sensitivity analysis, modeling changes in a variable against its effect; and value at risk (VaR) analysis, based on volatilities to calculate the chances of certain outcomes.

Let us look at a simple firm with earnings before interest and tax (EBIT) of 100, borrowings of 400 (all on a floating rate), an interest rate of 6% (as a base case), and a tax rate of 30%, and apply some of these techniques.

Evaluation 1: Sensitivity Analysis
A 1% move in interest rates has an effect of 4 (1% of 400) on the annual interest charge. This is not very helpful because there is no context for the effect.

Evaluation 2: Sensitivity Analysis
A table can be constructed to show the effect on earnings and interest cover (Table 1). In the table items in bold represent the base case, whereas other columns represent the sensitivities to this base case. Earnings are earnings after interest and tax.

Table 1. The effect of interest rate changes on earnings and interest cover

Interest rate	4.5%	5.0%	5.5%	6.0%	6.5%	7.0%	7.5%
EBIT	100.0	100.0	100.0	**100.0**	100.0	100.0	100.0
Interest	(18.0)	(20.0)	(22.0)	**(24.0)**	(26.0)	(28.0)	(30.0)
Tax	(24.6)	(24.0)	(23.4)	**(22.8)**	(22.2)	(21.6)	(21.0)
Earnings	57.4	56.0	54.6	**53.2**	51.8	50.4	49.0
Interest cover	5.56	5.00	4.55	**4.17**	3.85	3.57	3.33

Managing Interest Rate Risk

This is much more helpful, showing the effect on both earnings and interest cover. If the firm has an interest cover covenant of, say, 3.75, then the table shows a high risk of a breach, depending on how likely a rise in rates might be.

Evaluation 3: Sensitivity Analysis
Suppose now that EBIT displays volatility. We can construct a further table (Table 2) showing interest cover under variations in EBIT and the interest rate. Italic numerals indicate a covenant breach, and the number in bold is the base case described in Table 1.

Table 2. Interest cover under variations in EBIT and interest rate

Interest rate	4.5%	5.0%	5.5%	6.0%	6.5%	7.0%	7.5%
EBIT							
80	4.44	4.00	*3.64*	*3.33*	*3.08*	*2.86*	*2.67*
85	4.72	4.25	3.86	*3.54*	*3.27*	*3.04*	*2.83*
90	5.00	4.50	4.09	3.75	*3.46*	*3.21*	*3.00*
95	5.28	4.75	4.32	3.96	*3.65*	*3.39*	*3.17*
100	5.56	5.00	4.55	**4.17**	3.85	*3.57*	*3.33*
105	5.83	5.25	4.77	4.38	4.04	3.75	*3.50*
110	6.11	5.50	5.00	4.58	4.23	3.93	*3.67*
115	6.39	5.75	5.23	4.79	4.42	4.11	3.83
120	6.67	6.00	5.45	5.00	4.62	4.29	4.00

A drop of 5 in EBIT and a rise of 0.5% in interest rates will cause a breach, a clear risk factor for the firm. If a relationship between EBIT and interest rates can be established, then further conclusions could be drawn.

Sensitivity analysis does not show the probability of these changes, but if they are available—for example from a study of market volatility—a probability distribution for a covenant breach can easily be obtained.

Evaluation 4: VaR
Suppose that investigation of the assets and liabilities in the firm's pension scheme shows that the scheme has a deficit of 50. As an illustration, VaR might tell us that, based on the volatility of the long-term interest rates used to calculate liabilities, and taking into account that the scheme has some bond investments (in which value moves are opposite to liabilities), there is a 1 in 20 chance that the deficit will increase in the next year, because of interest rate changes alone, by 15 or more.

Interest rate risk inside a pension scheme (or other scheme for future employee benefits) often dwarfs interest rate risk inside the firm.

Evaluation should reveal where a firm is sensitive to interest rates. It could be:

- ▸ Earnings, perhaps where earnings per share (EPS) is an important issue.

Risk Management in an Uncertain World

- Cash flow.
- Interest cover ratios, perhaps because of financial covenants.
- Other ratios, such as those used by credit rating agencies.

Establishing a KPI and Response to the Risk

Evaluation should lead the firm to establish a KPI for interest rate risk. A good example of a KPI would be: *Interest cover to be greater than 3.75, on a 99% confidence basis, over an 18-month period.* This is better than using a simple interest cover ratio or a fixed/floating ratio as a KPI, because it speaks specifically about the risk to the firm.

The KPI should guide the response to the risk. Possible responses include:

- *Avoid:* It is hard to avoid interest rate risk.
- *Accept:* Simply accept the risk and take no further action. This may be suitable if there are no significant issues such as proximate financial covenants.
- *Accept and reduce:* It may be possible to reduce the risk through internal actions, such as reducing cash balances as far as possible to repay debt.
- *Accept and transfer:* Many market products are available that enable a firm to change the character of interest payments. This process is called hedging.

Establishing a Policy

The factors we have seen should be formalized in a policy, as should approaches to all risks. The policy should set out:

- The overall direction of the policy.
- How the risk is to be measured.
- Who has responsibility for the risk management.
- What procedures should be in place to control the risk.
- A framework for decision-making.
- The key performance indicator.
- A reporting mechanism to view the performance of the policy.

Tools Available to Transfer Interest Rate Risk

There are a large number of tools available for the transfer of interest rate risk (Table 3).

Table 3. The effect of interest rate changes on earnings and interest cover

Tool	Description	Comment
Forward rate agreement (FRA)	An FRA is a tool for fixing future interest rates (or unfixing them) over shorter periods, up to say 1–2 years.	A 3v6 FRA allows a firm to fix the three-month Libor (or other reference) rate in three months time. It is dealt over the counter (with banks).
Future	Futures have the same function as FRAs.	Futures are traded on an exchange, and thus have less flexibility.

Managing Interest Rate Risk

Tool	Description	Comment
Cap	A cap is an option instrument. The buyer of a cap pays a maximum interest rate over the life of the cap but enjoys lower rates as they come down. Caps have a premium.	Caps are usually dealt over the counter by firms, and the classic use is for a borrower to buy a cap that is higher than current interest rates, thus providing insurance for the borrower.
Floor	A floor is an option instrument. The buyer of a floor receives a minimum interest rate over the life of the floor but enjoys higher rates as they increase. Floors have a premium.	Floors are usually dealt over the counter by firms, and the classic use is for a depositor to buy a floor that is lower than current interest rates, thus providing insurance for the depositor.
Collar	A collar is a combination of a cap and a floor, thus providing a firm with a corridor of possible interest rates between a maximum and a minimum.	A borrower would buy a cap and sell a floor, usually over the counter, thus creating a "collar," or corridor, of rates.
Interest rate swap	An interest rate swap is probably the most widely used and popular risk transfer instrument in the field of interest rate risk. It changes the nature of a stream of interest payments from floating to fixed or vice versa.	Swaps (as they are usually called) are dealt over the counter and the market is large and (usually) deep. Terms of five to seven years are common with nonfinancial firms, although terms of 30 or more years are often used by pension schemes, reflecting their different maturity horizon.
Swaption	A swaption is an instrument where the buyer of a swaption has the right to enter into an interest rate swap at a particular rate, thus protecting the buyer against adverse movements in long-term rates, while allowing him/her to benefit from favorable moves.	Swaptions are not very popular with nonfinancial firms but might be used near the time of bond issues, for example.

Interest Rate Swap

This key instrument deserves a little more explanation. It is an instrument that, in its usual form, transforms one kind of interest stream into another, such as floating to fixed or fixed to floating. Each swap has two counterparties, and therefore in each swap one party pays fixed and receives floating, while the other party receives fixed and pays floating.

There are two classic uses of swaps by nonfinancial firms:

>> *A floating-rate borrower converts to a fixed rate.* In this case a borrower has floating-rate bank debt and carries out a pay-fixed swap, converting the debt to a fixed rate. This is shown diagrammatically in Figure 1. The two floating-rate streams cancel each other out for the borrower, leaving it to pay only a fixed-rate stream.

63

Risk Management in an Uncertain World

Figure 1. Floating-rate borrower uses swap to convert to a fixed rate

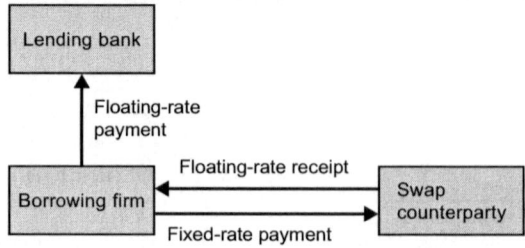

▸▸ *A fixed-rate borrower converts to a floating rate.* In this case a borrower has fixed-rate bond debt and undertakes a receive-fixed swap, converting the debt to a floating rate (Figure 2). The two fixed-rate streams cancel each other out for the borrower, leaving it to pay only a floating-rate stream.

Figure 2. Fixed-rate borrower uses swap to convert to a floating rate

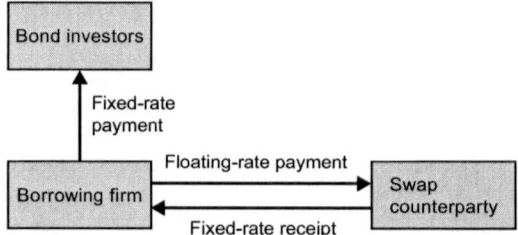

Let's suppose that our firm from above has responded to the risk of covenant breach by deciding to enter into a pay-fixed swap for 75% of its borrowing. It will pay 6% on the fixed-rate leg of the swap. The interest cover table we considered in Table 2 is now as shown in Table 4.

Table 4. Interest cover under variations in EBIT and interest rate for firm that pays fixed swap (see text for details)

Interest rate	4.5%	5.0%	5.5%	6.0%	6.5%	7.0%	7.5%
EBIT							
80	3.56	3.48	3.40	3.33	3.27	3.20	3.14
85	3.78	3.70	3.62	3.54	3.47	3.40	3.33
90	4.00	3.91	3.83	3.75	3.67	3.60	3.53
95	4.22	4.13	4.04	3.96	3.88	3.80	3.73
100	4.44	4.35	4.26	4.17	4.08	4.00	3.92
105	4.67	4.57	4.47	4.38	4.29	4.20	4.12
110	4.89	4.78	4.68	4.58	4.49	4.40	4.31
115	5.11	5.00	4.89	4.79	4.69	4.60	4.51
120	5.33	5.22	5.11	5.00	4.90	4.80	4.71

Managing Interest Rate Risk

The italicized cells (covenant breach) now cover the width of the table but are less deep. Our firm has a lower risk of a breach from interest rates alone but has increased the risk from a falling EBIT. As interest rates are believed to be more volatile than EBIT, the overall risk to our firm has been reduced through the transfer of risk.

Fixing Products versus Options

There is a key difference between interest-rate-fixing products (such as swaps) and options. A fixing instrument binds its user to the rate that is set when it is transacted. An option allows the buyer to walk away. So a firm taking out a pay-fixed swap, following which rates decline, is left paying the higher rates. The risk is thus transformed, rather than transferred. Exposure to rising rates has become an exposure to falling rates. Firms must be clear about this when establishing their response to risk.

Accordingly, option products may seem to be an ideal product to deal with interest rate risk, and for those prepared to pay, they can be. However, costs rise with two main factors:

- Time: The longer an option has until expiry, the higher the premium.
- Volatility: The higher the volatility in the underlying risk being hedged, the higher the premium.

Both these factors tend to deter firms from using options and, for the longer term, risk transfer-response interest rate swaps are usually the instrument of choice.

Conclusion

The effects of changes in interest rates on a firm can be complex, but techniques are available to evaluate and respond to any risks this presents. A clear reference back to business and financial strategy will put interest rate risk in its context, allow a suitable response, and help the firm to achieve its goals.

Making It Happen

- Assess how the firm is affected by changes in interest rates.
- Evaluate the risk according to the firm's strategy, using tools such as sensitivity analysis or VaR.
- Establish a key performance indicator for the risk.
- Choose whether to avoid or to accept the risk.
- If the choice is to accept, either:
  - accept and reduce; or
  - accept and transfer, such as with interest rate swaps or options.
- Make frequent reports to give feedback on the risk.

More Info

Books:
Buckley, Adrian. *Multinational Finance.* 5th ed. Harlow, UK: Pearson Education, 2004.

Chapman, Robert J. *Simple Tools and Techniques for Enterprise Risk Management.* Chichester, UK: Wiley, 2006.

Websites:
Association for Finance Professionals (AFP): www.afponline.org
Association of Corporate Treasurers (ACT): www.treasurers.org
National Association of Corporate Treasurers (NACT; US): www.nact.org

Risk to an Organization's Reputation

Understanding Reputation Risk and Its Importance

by Jenny Rayner

Abbey Consulting, Northwich, Cheshire, UK

This Chapter Covers

- Reputation is a critical intangible asset; it is an indicator of past performance and future prospects.
- Reputation is based on stakeholders' perceptions of whether their experience of a business matches their expectations.
- Knowing your major stakeholders, how they perceive you, and what they expect of you is vital in managing reputation risk.
- Everyone working for an organization bears some responsibility for upholding its reputation.
- Reputation risk is anything that could *impact* reputation—either negatively (threats) or positively (opportunities).
- Risks to reputation should be integrated into the business's enterprise risk management (ERM) framework so that they receive attention at the right level and appropriate actions are taken to manage them.

Introduction

Reputation is the single most valuable asset of most businesses today—albeit an intangible one. Hard-earned reputations can be surprisingly fragile in the globalized, technologically interconnected twenty-first century. The trust and confidence that underpin them can be irrevocably damaged by a momentary lapse of judgment or an inadvertent remark.

That is why understanding reputation risk has become a key focus for businesses in all sectors. It is now recognized that reputation risks need to be managed as actively and rigorously as other more quantifiable and tangible risks.

Reputation and Its Value

Reputation is an accumulation of perceptions and opinions about an organization that reside in the consciousness of its stakeholders.

An organization will enjoy a good reputation when its behavior and performance consistently meet or exceed the expectations of its stakeholders. Reputation will diminish if an organization's words and deeds are perceived as failing to meet stakeholder expectations, as illustrated by the reputation equation below.[1]

$$\text{Reputation} - \text{Experience} = \text{Expectations}$$

Reputation has intrinsic current value as an intangible asset. Although reputation will not appear as a discrete balance sheet item, it represents a significant

69

proportion of the difference between a business's market and book values (less any quantifiable intangibles such as licenses and trademarks). Since intangibles usually represent over 70% of market value, reputation is often a business's single greatest asset.

Reputation also plays a pivotal role in a business's future value by influencing stakeholder behavior and, hence, future earnings potential and prospects. A good or bad reputation can affect stakeholder decisions to maintain or relinquish their stake—be they investors, customers, suppliers, or employees. The "corporate halo" effect of a reputable business can help to differentiate products in a highly competitive sector, may allow premium pricing, and can be the ultimate deciding factor for a prospective purchaser of services. A strong reputation can help to attract and retain high-quality employees and can deter new competitors by acting as a barrier to market entry. Reputation can also shape the attitude of regulators, pressure groups, and the media towards a business and can affect its cost of capital.

Perhaps the greatest benefit of a good reputation is the buffer of goodwill it provides, which can enable a business to withstand future shocks. This "reputational capital," or "reputation equity," underpins stakeholder trust and confidence and can persuade stakeholders to give a business the benefit of the doubt and a second chance when the inevitable unforeseen crisis strikes.

Defining Reputation Risk

Reputation risk should be regarded as a generic term embracing the risks, from any source, that can *impact* reputation, and not as a category of risk in its own right. Regulatory noncompliance, loss of customer data, unethical employee behavior, or an unexpected profit warning can all damage reputation and stakeholder confidence.

Reputation risk is not only about downside threats, but also about upside opportunities. Climate change, for example, is a potential business threat, but many firms have spotted and exploited the flip-side opportunity for competitive advantage by developing green technologies and promoting themselves as environmentally friendly, thereby enhancing their reputation.

Reputation risk can therefore be defined as: "Any action, event, or situation that could adversely or beneficially impact an organization's reputation."

Identifying Reputation Risks

The most crucial stage of the reputation risk management process is *identifying* the factors that could impact reputation. Risks have to be recognized and understood before they can be managed. Considering the seven drivers of reputation is a useful starting point, as these are also fertile sources of threats and opportunity to reputation (see Figure 1).

Understanding Reputation Risk and Its Importance

Figure 1. The seven drivers of reputation. (*Source*: Rayner, 2003)

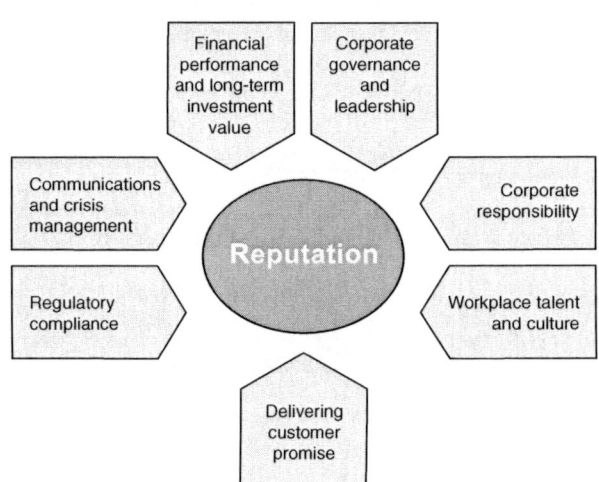

Businesses should consider not only the risks under their direct control, but also risks in the "extended enterprise" relating to suppliers, subcontractors, business partners, advisers, and other stakeholders. Could the values, business practices, or activities of its partners expose the business to reputation risk by association?

One way of approaching this is to consider the expectations of each major stakeholder group against the drivers of business reputation to develop a "heat map" of potential trouble spots and zones of opportunity. Major mismatches between expectations and experience can be analyzed to highlight areas where action is needed to bridge the gaps.

Asking the following questions may also help to uncover reputation risks:

- What newspaper headline about your business would you least (or most) like to see? What could trigger this?
- What could threaten your core business values or your license to operate? Such risks can seriously damage reputation and lead to an irreversible loss of stakeholder confidence.
- Could there be collateral risk arising from the activities of another player in your sector? If so, the reputation of your own business may be vulnerable and come under intense stakeholder scrutiny.
- Could reputation risk exposure arise from an acquisition, merger, or other portfolio change? A mismatch of values, ethos, culture, and standards resulting in inappropriate behavior could seriously damage reputation. Conversely, if the acquisition target enjoys a superior reputation, it could provide a competitive edge.

Risk Management in an Uncertain World

Evaluating, Responding To, Monitoring, and Reporting Risks

Once risks to reputation have been identified, they can be evaluated, appropriate risk responses developed, and the risks monitored and reported.

Risks to reputation can be *evaluated* in the usual way by considering the likelihood of the risk occurring and the impact if it does. The reputational impact of such risks should be considered explicitly, alongside financial or other impacts. This can be done by the use of a word model which explains reputational impact in a way that is relevant and meaningful for a given business. Table 1 provides an example of a four-point reputation impact scale that caters for both threats and opportunities.

Table 1. Sample reputation impact assessment criteria

Low	Moderate	High	Very high
Local complaint or recognition Minimal change in stakeholder confidence Impact lasting less than one month	Local media coverage Moderate change in stakeholder confidence Impact lasting between one and three months	National media coverage Significant change in stakeholder confidence Impact lasting more than three months Attracts regulator attention or comment	National headline/international media coverage Dramatic change in stakeholder confidence Impact lasting more than 12 months or irreversible Public censure or accolade by regulators

In assessing reputational impact, the view of relevant stakeholders should be considered to ensure that the impact is not underestimated. That is why understanding stakeholders and what they regard as current and emerging major issues lies at the heart of reputation risk management.

Reputational impact can sometimes be quantified in monetary terms—for example, expected reduced income resulting from loss of customers or license to operate; or impact on share price or on brand value. The true ultimate impact can be difficult to estimate as the immediate consequence may be only a relatively small financial penalty (for example, a fine for pollution). However, the event may, over time, have an insidious effect which erodes the business's reputation (for example, because of a perception that the business is not concerned about the environment).

Response plans should be developed to manage the more significant risks that present unacceptable exposure to the business. The gap between experience and expectation can be bridged by improving the business's performance or behavior and/or by influencing stakeholder expectations so they are more closely aligned with what the business can realistically deliver. As reputation is based on stakeholder perception, focused and clear communication to stakeholders is vital so that their perception will accurately reflect business reality.

A business may have done everything possible to anticipate and guard against reputational threats, but if a crisis strikes and the business response is inappropriate, its reputation may

Understanding Reputation Risk and Its Importance

still end up in tatters. Having an effective and well-rehearsed generic crisis management plan that can be quickly adapted and implemented to suit specific circumstances is therefore a key component of an effective reputation risk management strategy.

Once risks to reputation have been identified and responses agreed and implemented, the risks can be regularly *monitored* by management to ensure that responses are having the desired effect. Finally, the up-to-date status of the risks should be *reported* at the right level to inform decision-making and enable external disclosure to stakeholders.

Roles and Responsibilities

The board of a business is the ultimate custodian of a business's reputation. However, managing reputation risk successfully requires a team effort across the business from executive and nonexecutive directors, senior and middle managers, public relations staff, risk and audit professionals, and key business partners.

Everyone employed by and indirectly working for a business should be expected to uphold the business's values and bear some responsibility for spotting emerging risks that could impact reputation. The telltale signs of an imminent crisis are often missed because personnel are not risk-aware: a spate of customer complaints, safety near-misses or supplier nonconformance, a sudden rise in employee turnover, or pressure group activity. These can act as crucial early warning indicators which allow a business to take corrective action and avert disaster.

Case Study

Citigroup

In September 2004 the Financial Services Agency (FSA), Japan's bank regulator, ordered Citigroup to close its private banking business in the country following "serious violations" of Japanese banking laws. An FSA investigation found that inadequate local internal controls and lack of oversight from the United States had allowed large profits to be "amassed illegally." The bank had failed to prevent suspected money laundering and had misled customers about investment risk. The punishment meted out by the FSA was particularly severe as a previous inspection in 2001 had exposed similar compliance weaknesses, which Citigroup had not corrected.

Citigroup's then chief executive, Charles Prince, visited Japan in October 2004 in an attempt to repair the company's tarnished image. Bowing, he apologized for the activities of his senior staff, saying that they had put "short-term profits ahead of the bank's long-term reputation." He pledged to improve oversight, change the management structure, increase employee training on local regulations, and set up an independent committee to monitor progress. He said: "Under my leadership, lack of compliance and inappropriate behavior simply will not be tolerated and we will take direct action to ensure that proper standards are upheld and that these problems do not reoccur."

That same month French retailer Carrefour fired Citigroup as a financial adviser on the sale of its Japanese operations to prevent its own reputation from being tarnished by association.

Conclusion

A good reputation hinges on a business living the values it claims to espouse and delivering consistently on the promise to its stakeholders. Being "authentic," being "the real thing," has never been so important. Pursuing short-term gain at the expense of long-term business reputation and stakeholder interests is no longer acceptable practice.

Successfully managing reputation risk is both an inside-out and an outside-in challenge. The inside-out component requires business leaders to establish an appropriate vision, values, and strategic goals that will guide actions and behaviors throughout the organization. The outside-in component requires the business to scan the external environment continuously and canvass stakeholder opinion to ensure it is on a track that will secure the continuing support, trust, and confidence of its stakeholders.

Active and systematic management of the risks to reputation can help to ensure that perception is aligned with reality and that stakeholder experience matches expectations. Only in this way can a business build, safeguard, and enhance a reputation that will be sustainable in the long term.

Making It Happen

The key components of reputation risk management are:

- Clear and well-communicated business vision, values, and strategy that set the right ethical and stakeholder-aware tone for the business.
- Supporting policies and codes of conduct that guide employee behavior and decision-making so that goals are achieved in accordance with business values.
- Extension of the business's values and relevant policies to key partners in the supply chain.
- Dialogue and engagement to track the changing perceptions, requirements, and expectations of major stakeholders continuously.
- An effective enterprise-wide risk management system that identifies, assesses, responds to, monitors, and reports on threats and opportunities to reputation.
- A culture in which employees are risk-aware, are encouraged to be vigilant, raise concerns, highlight opportunities, and act as reputational ambassadors for the business.
- Transparent communications that meet stakeholder needs and build trust and confidence.
- Robust and well-rehearsed crisis management arrangements.

More Info

Books:
Atkins, Derek, Ian Bates, and Lyn Drennan. *Reputational Risk: Responsibility Without Control? A Question of Trust*. London: Financial World Publishing, 2006.
Fombrun, Charles J., and Cees B. M. van Riel. *Fame and Fortune: How Successful Companies Build Winning Reputations*. Upper Saddle River, NJ: FT Prentice Hall, 2003.
Larkin, Judy. *Strategic Reputation Risk Management*. Basingstoke, UK: Palgrave Macmillan, 2003.

Understanding Reputation Risk and Its Importance

Rayner, Jenny. *Managing Reputational Risk: Curbing Threats, Leveraging Opportunities.* Chichester, UK: Wiley, 2003.

Article:

See articles in *The Geneva Papers on Risk and Insurance Issues and Practice* 31:3 (July 2006). Online at: tinyurl.com/8y9lksb

Reports:

Coutts and Company. "Face value: Your reputation as a business asset." London: Coutts and Company, 2008.

Economist Intelligence Unit. "Reputation: Risk of risks." White paper. 2005.

Resnick, Jeffrey T. "Reputational risk management: A framework for safeguarding your organization's primary intangible asset." Opinion Research Corporation, 2006.

Websites:

John Madejski Centre for Reputation, Henley Business School, University of Reading: tinyurl.com/7l5zqba

Reputation Institute: www.reputationinstitute.com

Notes

1. Oonagh Mary Harpur in Chapter B4 of *Corporate Social Responsibility Monitor*. London: Gee Publishing, 2002.

Managing Reputational Risk: Business Journalism

by Jonathan Silberstein-Loeb

Centre for Corporate Reputation, Saïd Business School, University of Oxford, UK

> **This Chapter Covers**
>
> » How credible commitments can help to mitigate reputational risk.
> » Establishing credible commitment, and trust, requires a better understanding of the incentives of journalists, as well as the perceptions of business, both of which are explored in this chapter.
> » The consequences of these differing incentives for building trust between journalists and businesses and for company strategy are explored in the final section.

Introduction

It is taken as given that journalism affects corporate reputation. To understand better how journalism about businesses is written and what can be done to contend with the reputational risk that it may present, this chapter seeks to explain how business practitioners and business journalists perceive each other and interact. It is the central argument of this chapter that establishing mechanisms for credible commitment helps to mitigate reputational risk. As there is no formal mechanism for establishing credible commitment between business representatives and business journalists, doing so depends on their respective incentives and corresponding informal constraints, such as career advancement and the loss of reputation. Insofar as journalists must have sources, and sources must communicate with their stakeholders, journalists and corporate decision-makers rely equally on each other, and therefore may be hostages of one another in instances of repeated interaction. The implicit recognition that both parties are beholden to each other may be the most effective mechanism of credible commitment. Behavior that demonstrates an awareness of this codependence helps to establish trust, which is central to effective media relations.[1] Trusting relationships facilitate an understanding of the incentives that undergird credible commitments, which helps to make them reinforcing. Recognizing the importance of credible commitments has clear consequences for the way in which companies develop communication strategies.

Much of what follows is predicated on the preliminary results of an international study of business journalism and corporate reputation that the Oxford University Centre for Corporate Reputation is conducting. This chapter relies on evidence gathered as part of this study in more than 80 informal, off-the-record interviews conducted in England with journalists from major national dailies that publish business sections and with business leaders (see Appendix 1).

The Incentives of Journalists
For Whom Do Journalists Write?
It might reasonably be expected that the audience for business journalism would affect journalists' incentives, but journalists have, at best, an imperfect impression of the reader

for whom they write, or who reads their articles. "The general reader" is a phrase that journalists frequently use, but it lacks a clear definition. There is considerable difference among writers at each publication, which makes for little differentiation between publications, although the *Financial Times* is tailored toward the investment community more than the *Sunday Times*. Some business journalists claim to write for the person on the Clapham omnibus, others for the clerk in the City. The public is largely financially illiterate, and journalists are obliged to explain the basics to the general reader, but this duty does not undermine the value that business news holds for specialists. Those business professionals who read the business sections—and most do—do so less for insight than for context.

What Motivates Journalists?
Journalists' motivations affect their incentives. In a survey of American journalists that has been conducted four times over the past 30 years, Weaver and Wilhoit (1996) identified three principal functions among journalists: disseminators of information; interpreters of events; and adversaries of business and government. Throughout the survey period, the dominant professional role of journalists has been "interpretive"— that is, to provide analysis and interpretation of complex issues. Since the 1970s, it has always been the case that only a very small minority of the sample felt that the adversary role was important. The percentage of journalists surveyed who believed that it was extremely important for the mass media to "be an adversary of business by being constantly skeptical of their actions" has always been low, even in the wake of dozens of scandals involving large corporations. The percentage of journalists who believed the adversarial function to be extremely important was greatest among those working for news magazines, and least among those working for radio. Although these data pertain to American journalism, cultural similarities between Britain and the United States, and responses from journalists interviewed in London, provide reasons to suspect that a like study carried out in the United Kingdom would produce similar results.

The motivation of business journalists in part derives from the way in which they perceive the companies about which they report. Opinions vary from journalist to journalist, and journalists change their minds over time. When they believe they are on to a hot story, their approach may be adversarial; when they are covering a diary event, it may be interpretive. The journalists who were interviewed evinced as much a desire simply to "understand how the world works" as to serve an interpretive or watchdog function, although many professed themselves to be skeptical of authority. Regardless of their perceived role, most journalists interviewed believe that they can hold business to account when required. It is broadly the case, and common sense, that business journalists at the *Financial Times* tend to see themselves as serving more of an interpretive function, whereas journalists at *The Guardian* tend to see themselves more as watchdogs of business. Journalists at the *Daily Mail* were also more likely to think it their job to hold business to account. Journalists at other publications are either indifferent, or opinions among them vary so considerably that any characterization by publication is impossible.

How Do Journalists Interact with Business Sources?
Regulations respecting the disclosure requirements imposed on businesses significantly affect the quantity and quality of information available to journalists, but the material

Managing Reputational Risk: Business Journalism

divulged in disclosure documents is in the public domain. Such material is unlikely to be newsworthy, especially now that the Internet makes it so easy to search for and share this information. Journalists require scoops, and for this reason, contacts are critical to all journalists, regardless of publication. How journalists develop these contacts varies according to publication, and to style of writing. Breakingviews.com and *The Economist*, for example, as well as columnists at the dailies, need not worry about quoting sources on-the-record. If discussions are *de facto* off-the-record, then journalists have an easier time getting information. The perception of a particular publication, and its perceived audience, among sources is also influential: people in the City are more likely to find time for a chat with journalists from the *Financial Times* than they are with journalists from *The Guardian*. Consequently, those journalists who see themselves as watchdogs find it more difficult to cultivate the sources they require to fulfill this function.

There is irony here that bespeaks a fundamental problem in all forms of journalism. Readers of *The Guardian*, and similar publications, are less interested than readers of the *Financial Times* in the daily operations of big business. Journalists who see themselves as adversaries of business tend to work for publications that circulate among readers with a dislike for big business, or at least only a secondary interest in it. Given their readership, adversarial journalists are therefore less likely than journalists working at business-friendly publications to contact, or be contacted by, the very people whom they seek to hold to account. Journalists waging crusades tend to operate on the margins, and to rely on workers in the third sector for information, rather than the business decision-makers, on whom journalists at business-orientated publications tend to rely. Further, publications tailored for business, such as Breakingviews.com, *The Economist*, the *Financial Times*, and the *Wall Street Journal*, have been more successful than general publications at charging for their online content, not least because the people who read these publications perceive them to be important to their careers. Publications tailored for business audiences have more resources to devote to serving a watchdog function, but perhaps less incentive to do so. The fact that News Corporation owns Dow Jones, which covers day-to-day events, enables journalists at the *Wall Street Journal Europe*, which News Corporation also owns, to devote more time to in-depth reporting, and yet comparatively few journalists at the *Wall Street Journal Europe* see themselves as watchdogs of business.

Regardless of the way in which journalists perceive their role, it is broadly the case that they are concerned to advance their careers—and obtaining scoops, or exclusive news, is an effective way for them to do so. Although journalists may obtain exclusive news either through investigation, which is typically adversarial, or through sources, which rely on relationships, in practice the two often overlap: sources provide leads, which require further investigation, which requires sources. The reality is that journalists rely on their contacts and the strength of their relationships with sources to do their jobs effectively. Whether journalists perceive their function to be interpretive or adversarial will determine the extent to which they are willing to bite the hands that feed them. In an age of WikiLeaks and whistleblowing, journalists will always find someone with a story, but those journalists with a proclivity for provocation are likely to find that their pool of contacts dries up quickly. All journalists must constantly query whether flouting or favoring their sources best serves their interests. If journalists are unwilling to cross their sources, they risk being mere conduits.

Journalists Have a Code of Behavior

To further their careers and to obtain quality contacts, journalists must maintain their credibility, which is achieved through accuracy and consistency. Relationships with sources are built up over years and are based on trust. Maintaining these relationships requires journalists to be accurate and honest. Unsurprisingly, all journalists profess to uphold these standards and to behave professionally. Assuming journalists are committed to doing their jobs well, which may be a large assumption, they will seek to obtain access to credible and authoritative sources, which typically means bypassing PR in favor of speaking directly with executives. They will also attempt to corroborate the facts of a story through a process of triangulation.

The relationship between journalist and source can be a game of cat and mouse, one of tit-for-tat, of "you scratch my back, I'll scratch yours," or a continuous negotiation. Journalists know, and business decision-makers confirm, that without prior relationships journalists will be directed first to PR, although most journalists hope to speak to C-suite executives on every story. Even then, to talk to a company that is not PR-trained is rare. Some sources are more forthcoming than others. Most of the journalists interviewed believe that external PR professionals are better sources than internal communications officers, but it is occasionally the case that company employees are especially well briefed and agency employees are uninformed. On the one hand, to the extent that journalists perceive PR to be a hindrance that must be circumvented, the presence of PR may radicalize journalists; on the other hand, if a PR professional, or any source, is discovered to have lied to a journalist, they will lose their credibility as a source. Among business journalists, there exist the "good" and "smart" PR professionals who understand this game, and the "bad" PR who "don't get it." A process of name-and-shame in newsrooms generates and promulgates a source's reputation.

The Incentives of Business Representatives

Most companies proactively undertake to conduct, in conjunction with an external financial public relations agency, a series of strategic, staged formal meetings with journalists throughout the year that coincide with annual announcements. Interactions between companies and journalists more frequently take place in informal environments, which, from a corporate perspective, are less orchestrated and strategic, and which consequently have potentially greater reputational repercussions.

Willingness of Business to Engage with Journalists

Willingness to engage with the media varies, but some form of interaction is necessary and important. Saying "no comment" may make it difficult for journalists to write a story, but more often it tends to annoy them, and it rarely prevents them from writing. Few C-suite executives court business journalists, but nearly all are convinced of the importance and beneficial effects of trusting relationships. As a general rule, CEOs tend to interact more with the media than with nonexecutives. Chairpersons only get involved when there is a strategic event, such as a takeover bid or a governance issue. Much of the interaction with journalists tends to focus around the reporting of company results, but executives make efforts to have regular contact with journalists to provide background information. Much of this interaction is conducted under the guise of helping journalists to understand the business, explaining decisions made, or making sure they get the facts of the story straight.

Managing Reputational Risk: Business Journalism

The Importance of Trusting Relationships

Trusting relationships may help CEOs to avoid negative coverage. Creating such relationships with journalists entails interactions outside work-related environments. These informal interactions, typically conducted off-the-record, may be informative for journalists and executives. CEOs and other executives are convinced that even if they have lousy news to give, they receive more balanced treatment from journalists they know. A "professional relationship," according to many business leaders, helps them to promulgate messages that kill adverse rumors and gossip. Long-term relationships make journalists more receptive to the views businesses put across. It can be dangerous to court journalists, but once a trusting relationship is established, journalists will seek to maintain it through good behavior to ensure continued access to a reliable source. Developing effective relationships with journalists requires honest discussion and exchange. Trade in information does not characterize such relationships. Journalists may grant businesses favorable treatment on negative stories in exchange for off-the-record comments, but this a perilous strategy to pursue. Sustainable, effective relationships rely on a modicum of respect and quite a lot of patronizing. Trust is critical, and so is honesty.

Although personal contact creates more credible relations, evidence suggests that the Internet has made the relationship between journalists and corporations more mediated. The personal communication that previously followed the release of statements—and which allowed for clarification—has ceased, making public statements more important. The focus of media relations is now more on preparation than on relationship building. This shift also reflects the fact that traditional media are less important than they once were. Creating a case for a company that meets the needs of all its stakeholders requires waging battles on multiple fronts, and not just in the *Financial Times*. Media fragmentation also makes relationships between journalists and sources less manageable. The permanency, accessibility, and transportability of information on the Internet means that companies must have a clear message.

Creating Credible Commitments to Mitigate Reputational Risk

Journalists, whether they perceive themselves to be watchdogs or not, require information. To obtain this information, and to do their jobs effectively, journalists must gain the trust of their sources. Corporate decision-makers likewise have an incentive to gain the trust of journalists so that they may effectively communicate important information to the market. At the upper echelons of business and business journalism in London, theory and practice converge more often than not. There is an implicit recognition that both parties require each other.

Decision-makers have information at their disposal that journalists do not, and consequently they are obliged to convey this information to journalists frankly and in such a way that the journalists are satisfied that there has been full disclosure. The uneven distribution of information and the burdensome obligation of disclosure make mistrust and miscommunication likely. Journalists, by dint of their information deficit, are likely to be suspicious of their sources. Decision-makers, confronting such suspicion, are likely to be convinced that journalists are always looking for dirt. If a company is conducted in a strategically sound manner and within the law, it has little to worry about from the media. (Even if it is conducted in an unsound manner, recent history suggests that journalists may be slow to cotton on.) A consistently sound strategy achieves favorable results. Profits are facts on the ground with which it is difficult for even the most biased journalist to argue. If the business is performing poorly, it is only fair, and of service to the market, that this be reflected in the papers.

Yet, the blame game is subjective, and journalists are fallible, so it is reasonable that in the court of public opinion businesses should have advocates as they do in courts of chancery. Even the good have need of an advocate. Although the rules of the game that dictate the behavior of advocates in the court of public opinion are by necessity unwritten, and more informal than those regulations under which barristers labor, adherence to them is as necessary for the orderly distribution of justice, and flouting them is just as punishable. These are the rules that apply to the construction and maintenance of trusting relationships.

Trusting relationships diminish the prospect that rogue journalists will undeservedly lash out at corporations and their representatives. As with capital, during good times business leaders should seek to generate favorable relationships that are resistant to change during less fortuitous periods in the life of the organization. Journalists seek to build trusting relationships by obtaining a reputation for credibility, which is obtained through professionalism. Being credible does not mean being sympathetic to sources—it means being fair. For decision-makers to convey their opinions effectively to journalists, they too must be seen as being credible. The best way to achieve credibility with journalists is for business decision-makers to be open and honest. Of course, journalists may sensationalize a story to augment their own career or to sell newspapers, just as companies may employ strategies of obfuscation and spin, but for journalists opportunistic behavior undermines credibility, and for corporations spin is at best a fallible prophylactic.

A Continuous Process of Negotiation
It is useful to imagine relationships between journalists and corporate decision-makers as a continuous process of negotiation. In an insightful article, Charron (1989) observed that these relationships are "at once cooperative and fraught with conflict." The dual nature of these relationships, wrote Charron, "implies a double negotiation: over the exchange of resources, and over the rules regulating this exchange." Cooperation results from mutual interdependence of the players for resources. Each actor's dependence on the other varies according to the possibility of alternatives. A more adversarial dimension arises from attempts by either journalists or sources to manage this relationship. Charron concluded that collaboration and accommodation are the best strategies for sources, although these strategies entail risks, whereas journalists, when their interests are not directly at stake, will tend to adopt a strategy of avoidance rather than accommodation.

For many years PR professionals have known that "spin" is self-defeating. It is widely recognized that nuance, and presenting both sides of the story, is critical to maintaining the credibility of PR. In the upper echelons of PR in England these enlightened views hold sway. When possible, PR professionals ought to provide journalists with reliable information and facilitate their access to C-suite executives. All journalists recognize that sources will put their version of the story across, and they regard this as reasonable, but anything less than collaboration and accommodation is likely to incur the wrath of journalists and do a disservice to public relations. The power to improve media relations, and to augment corporate reputation, therefore lies with corporations. In the blame game, corporations are both more permanent and more manageable players than individual journalists, whose personality and practice is less readily determined or constrained. Mechanisms, such as professional codes of conduct, however fallible they may be, already exist to dissuade journalists from opportunistic behavior. Any further restrictions on journalistic activity would have an adverse effect on the social benefit they provide to the public. The onus to improve relations lies with business.

PR Should Play a Role in Strategy

Businesses ought to avoid putting those that communicate on their behalf in a position in which they are obligated to jeopardize their relationships with journalists. Instead, the PR function should play a more influential role in developing corporate strategy so that company behavior is justifiable to journalists, and so that sources are able to retain credibility with them. So long as businesses do not oblige sources to deceive journalists, media relations will be predicated on trust, which enables businesses to convey effectively the rationale behind strategic decisions.

During the past decade, there have been indications of a greater appreciation of the role and value of the PR function in management, but it is still seen by many as foremost a communications role. Corporate decision-makers are concerned about how actions will be portrayed on the cover of the *Daily Mail*, but there is little direct connection between PR and strategy. In part, this gradual increase in the managerial role of PR is due to the growing importance of investor relations. As Davis ("Public relations, business news and the reproduction of corporate elite power," 2000) has observed, publicly quoted companies periodically devote resources to government policy-making and institutional regulation, but they are more concerned with institutional shareholders and the wider business community. Such concerns translate into an emphasis on communications with investment analysts, business leaders, and business journalists.

Given the influence that the media may exercise on company reputation, executives must weigh the gains to be had from strategic actions against the losses to be expected from adverse media commentary. If the executive is convinced that the profits to be gained from a particular strategic action outweigh the potential loss to profits incurred from negative media coverage, the executive has a responsibility to carry out the action. If, however, the response of the media may be so adverse as to cause damage in excess of the benefit derived from strategic action, the executive has an equally powerful but opposite obligation to avoid the strategy, to modify it, or to explain it to the media in such a way as to limit losses when the strategy is carried out. The PR function should help in making this calculation. This is not to say that PR ought to be the conscience of the corporation; rather, PR professionals should bring to bear on the process of strategic planning the perspective and interest of different stakeholders whose support is critical to profitability.

When relationships between corporations and business journalists function effectively, they generate transparency and the amount of information about a company available to the media may consequently increase. Transparency enables the market to know what a company does, and why. All things being equal, a greater quantity of information will better ensure that market perceptions parallel corporate behavior. Transparency may compel companies to adhere to social norms and behave responsibly without sacrificing profitability. Having increased the quantity of information available to the media, and by extension to the public, corporations will have more to manage, but this outcome is conducive to the effective function of markets, which is necessary for economic growth generally and corporate growth specifically.

Risk Management in an Uncertain World

> **Summary and Further Steps**
> - Implicit recognition that journalists and corporate decision-makers require each other leads to a credible commitment that builds trust.
> - Closer cooperation between the managerial and communications functions within firms helps to facilitate trusting relationships with journalists, and consequently to mitigate reputational risk.

> **More Info**
>
> **Book:**
> Weaver, David H., and G. Cleveland Wilhoit. *The American Journalist in the 1990s: U.S. News People at the End of an Era*. Mawah, NJ: Lawrence Elbaum Associates, 1996.
>
> **Articles:**
> Charron, Jean. "Relations between journalists and public relations practitioners: Cooperation, conflict and negotiation." *Canadian Journal of Communication* 14:2 (1989): 41–54. Online at: tinyurl.com/6fg78qd [PDF].
>
> Davis, Aeron. "Public relations, news production and changing patterns of source access in the British national media." *Media, Culture & Society* 22:1 (January 2000): 39–59. Online at: dx.doi.org/10.1177/016344300022001003
>
> Davis, Aeron. "Public relations, business news and the reproduction of corporate elite power." *Journalism* 1:3 (December 2000): 282–304. Online at: dx.doi.org/10.1177/146488490000100301

Appendix 1
Interviewees for Oxford University Centre for Corporate Reputation study of business journalism and corporate reputation

Newspapers and journals	Business leaders	
BreakingViews.com	Torsten Altmann	Carol Leonard
Daily Mail	Norman Askew	David Mansfield
The Economist	John Barton	Peter Morgan
Financial Times	Alex Brog	John Peace
The Guardian	Roger Carr	Roger Parry
The Independent	Peter Cawdron	Sir Ian Prosser
The Times	Stuart Chambers	Michael Rake
Sunday Times	Doug Daft	Don Robert
Wall Street Journal Europe	Terry Duddy	Robin Saunders
	Steve Easterbrook	Oliver Stocken
	Andrew Grant	Robert Swannell
	Andy Hornby	John Tiner
	Lady Barbara Judge	David Tyler
	Frederick Kempe	Lucas Van Praag
	Roddy Kennedy	Sarah Weller
	John Kingman	Gerhard Zeiler

Note

1. I am grateful to Dr Paolo Campana for help on this point.

The Cost of Reputation: The Impact of Events on a Company's Financial Performance
by Daniel Diermeier
Kellogg School of Management, Northwestern University, Evanston, Illinois, USA

This Chapter Covers

- Reputational crises have a significant impact on a company's valuation.
- They can be triggered by any business activity and do not necessarily reflect lapses in a company's ethics or integrity.
- Both the frequency and the magnitude of such events is increasing.
- The underlying factors that drive these developments are likely to increase in importance.
- Since reputational risk cannot be hedged or "outsourced," companies need to develop effective reputation management capabilities.
- Such capabilities consist of an integrated reputation management system and its core components: (1) mindset, (2) processes, and (3) values and culture.
- A reputation management process consists of a decision-making system and an intelligence system.

Introduction

CEOs and board members routinely list reputation as one the company's most valuable assets. Yet every month a new reputational disaster makes the headlines, destroying shareholder value and trust with customers and other stakeholders. During the last year, leading companies such as Toyota, Goldman Sachs, BP, Johnson & Johnson, and HP battled severe reputational crises. In all cases, financial markets punished the companies, leading to a severe and sustained erosion of their market values. In many cases, reputational damage is followed by lawsuits, public hearings, investigations, and regulatory actions.

In contrast to the scandals related to Enron, WorldCom, and Arthur Andersen a decade earlier, these crises are not limited to a specific domain (accounting practices and standards, especially with "new economy" firms) or caused by a dramatic increase in blatantly unethical or illegal activities. Rather, the involved companies were all category leaders, some with iconic status in their respective industries, and the issues involved ranged from quality and safety to disclosure and (alleged) executive misconduct.

The increase in the frequency and impact of reputational issues suggests that more fundamental shifts are occurring in the business environment and that companies are unprepared for dealing with them. What companies lack is an effective reputation management capability in the presence of increasing reputational risk. Too often, reputation management is considered a (sub)function of corporate communication and isolated from business decisions. Rather, companies need to adopt a strategic approach that treats reputational challenges as understandable and even predictable.

As a result, companies should manage their reputation like any other major business challenge: based on principled leadership and supported by sophisticated processes and capabilities that are integrated with the company's business strategy and culture.

> **Case Study**
>
> ### Bausch & Lomb
> Markets do not always properly adjust to reputational risk. One such example is Bausch & Lomb, a producer of soft contact lenses and lens care products. On April 10, 2006, the US Centers for Disease Control and Prevention linked a surge in potentially blinding fungal infections with Bausch & Lomb's new ReNu contact lens solution. As a result, Bausch & Lomb's stock price dropped from a closing price of US$57.67 on Friday, April 7, to US$45.61 on Wednesday, April 12. The company was heavily criticized for its handling of the crisis and the depressed stock price persisted. Bausch & Lomb subsequently experienced accounting restatements and was acquired by the private equity firm Warburg Pincus.
>
> Remarkably, the link between the infections and ReNu, however, had been uncovered almost two months earlier, on February 22, in a public announcement by Singapore's Ministry of Health. (Bausch & Lomb subsequently withdrew the ReNu solution from its markets in Singapore and Hong Kong). The government announcement had been reported in the region's major newspapers, but had not been covered in the United States. Bausch & Lomb's stock lost a mere 3% from a closing price of US$71.51 on February 21 to US$69.40 on February 23, and quickly recovered. In other words, financial markets ignored the early warning signs.

The Cost of Reputational Crises
Severe erosion of shareholder value is common during reputational crises. During its recent crisis triggered by the sudden acceleration issue, Toyota's stock price dropped by as much as 24%, wiping out about US$33 billion in shareholder value, close to the total market value of Time Warner. In its battle with the US government in the aftermath of the 2008–09 financial crisis, Goldman Sachs lost US$24 billion of its market capitalization, a 26% drop in share price that exceeded the entire value of American Express. During the BP oil spill disaster in the Gulf of Mexico, BP's stock was almost cut in half, the equivalent of about US$90 billion in shareholder value, more than the market value of Procter & Gamble.

In some cases the drop in stock value is temporary, in other cases permanent. Much depends on how the companies handle the aftermath of crisis and commit to fixing the underlying business issue rather than engaging in shallow PR exercises. Toyota, for example, commenced a global quality improvement initiative that involved cultural and process changes at every level of the company.

The Problem of Measurement
In general, the impact of reputational events is difficult to quantify. Many existing studies point to a correlation between superior corporate reputations and financial performance. In most studies, corporate reputation is measured using lists from Fortune's Most Admired Companies.[1] There are various problems with this approach.

First, there is little change in the list membership. This may constitute evidence that reputation is persistent, but may also point to a metric that is too coarse to detect changes in a company's reputation on a smaller timescale, by region or product. Second, both inclusion in the list of most-admired companies and superior financial performance may be consequences of an underlying third characteristic that drives them both, such as "good management."

Studies that show the positive impact of reputation at the more general, macro-level are of limited use to managers as they typically fail to identify underlying factors driving a company's reputation. Given the difficulty with such macro-level studies, a better understanding of the processes that shape perceptions at the micro-level is desirable.

Reputation as a Bank Account

One common view of reputation is that it serves as a "trust bank account" or "buffer." The idea is that, through their actions, companies make "deposits" into a "trust account" that generates goodwill with their various stakeholders and constituencies. In case of a crisis, companies then are able to make a "withdrawal" from this bank account that helps to at least partially isolate them from the impact.

There is some evidence for this claim in the context of corporate social responsibility (CSR). Recent research suggests that in the case of a product recall, for example, companies with high CSR ratings lose on average US$600 million less in firm market value than companies with low ratings. However, other evidence suggests that, in a crisis context, previous "good deeds" are swamped by current actions (good or bad). Trust, it seems, can't easily be deposited; it must be earned.

Perception Drivers

To understand these processes in more detail, a rigorous approach using controlled experiments is helpful. The approach is similar to the market research studies that are conducted when companies design or evaluate brands. In such studies, subjects read vignettes of fictitious newspaper articles that describe a crisis. For example, a food manufacturer may be accused of using a potentially harmful food additive in order to increase the shelf life of its products, or the company may be involved in a nasty sexual harassment lawsuit. In addition to this background information, subjects are also provided with the fictitious company's responses, which range from engaged and caring to dismissive.

While a "trust bank account" (here measured by past good deeds) does have some effect, companies' current actions have a greater effect on shaping attitudes toward the company in question, as well as apparently unrelated issues such as the aesthetic evaluation of logos or product design.

Even more strikingly, when subjects were asked to evaluate the taste of a product (e.g. bottled water) from a company with a low "trust bank account," they rated the taste lower and drank less water. Corporate executives are only partially aware of these effects. When asked to predict how public attitudes would be affected by these same scenarios, corporate executives correctly predicted that an engaged response would be viewed more favorably than a defensive response, but they were overly optimistic about the public's ability to refrain from forming opinions when the company offered "no comment."

In sum, corporate reputation strategies have direct and measurable effects on the evaluation of core brand attributes. They can affect overall customer perceptions, evaluation of corporate logos, and even opinions of product taste and levels of consumption. Corporate executives are largely unaware of such effects.

Reputation as a Currency

These findings can be reconciled if we think of reputation less as a bank account and more as a currency. Currencies act as multiplies (not "linearly" as the bank account suggests). That is, if a currency is strong, purchasing power increases for all sorts of goods.

So, what does this mean in the context of reputation? The idea is that the same statement carries more weight and has more impact if it comes from a company with a good reputation. Or alternatively, a company with a strong reputation will need to do less than a company with a weak reputation to achieve the same effect.

Some evidence from the 2011 Edelman Trust Barometer[2] supports this intuition (Table 1).

Table 1. Effect of company trust. (*Source*: 2011 Edelman Trust Barometer)

Company trust	Will believe negative information if hearing it 1–2 times	Will believe positive information if hearing it 1–2 times
High	25%	51%
Low	57%	15%

In case of "low trust" we find the well-known negativity bias. A negative message has roughly four times as much impact as a positive message. But, strikingly, the situation is reversed in the "high trust" case, where a negative message has only half as much impact as a positive one. So, by moving from a high to a low trust "currency" a company gets an eight-fold increase of positive information impact—not bad.

Still, these results are to be treated with some caution. Ideally, one would want to replicate this behavior in a controlled laboratory setting. Yet this phenomenon would explain the "Teflon" characteristic of some companies, such as Apple, as witnessed in the limited impact of recent issues such as the dropped calls/antenna problem in the iPhone 4 on Apple's reputation. Surveys consistently put the high-tech industry at the top of the list of most-trusted industries.

Reputation Management Capabilities

The difficulty in exactly quantifying the impact of reputational events implies that reputational risk cannot easily be hedged or insured against. It therefore must be managed. This means that companies need to build reputation management capabilities appropriate to their level of risk.

The Cost of Reputation

A company's reputation needs to be managed *actively*. Good business practices and ethical conduct are necessary, but they alone are not sufficient for successful reputation management. That responsibility should lie with business leaders. It should not be delegated to specialists such as lawyers or public relations experts, even though such experts play a valuable role in the reputation management process. Integration with business practices is necessary as reputational challenges arise out of a specific business context. In many cases, the most effective way to manage reputational risk is to improve the capabilities of business leaders (supported by well-designed processes) rather than adding another corporate layer.

Reputational challenges can arise from any area of day-to-day decision making, but executives tend to make decisions without consideration for the reputational impact. The key skill for business leaders is the ability to maintain an external perspective throughout decision-making processes and incorporate this perspective into the design of business decisions, e.g. the launch of a new product and its market-entry strategy. Companies need to understand that their decisions are creating a record today that will serve as the basis for their story tomorrow. Assessing reputational risk requires anticipating what a reputational crisis would look like and then taking proactive steps to prevent and prepare.

During a reputational crisis, the spotlight will not only be on the company's *current* actions, such as how the CEO answers questions and what the company will do to fix the problem, but also on its past actions. Reporters will ask when the company first knew about the problem, or why management didn't do more to fix it. The thought process behind each past decision can be brought out into the public arena and questioned. These past actions and decisions are now part of the record and cannot be changed. Even actions that looked reasonable at the time may wither under scrutiny from a hostile audience in a crisis context after any negative consequences come to light.

After the Gulf of Mexico oil spill, every minute decision that BP made concerning its safety processes took on disproportionate significance, leading to severe criticism of the company. And when Toyota had to recall its cars, commentators quickly alleged that its aggressive growth strategy had sacrificed quality and safety.

An easy way to improve decision-making is the *Wall Street Journal* test, which suggests that decision-makers should ask themselves whether they would be proud if a decision were *accurately* reported on the front page of the *Wall Street Journal*. This test evocatively captures the idea that a decision may look different once it comes under public scrutiny.

Such approaches help to transition from a crisis management mindset to a risk management one. Taking reputational risk seriously does not necessarily mean refraining from giving the green light to decisions that carry some reputational risk. Rather, the goal of proactive reputation management is to identify possible risks and mitigate them through *current* actions to reach an acceptable balance level of risk and control.

Reputation Management Processes

Who should own reputation management? Many executives answer: everyone. That sounds reasonable enough, but it is easy for things that are owned by everybody to actually be owned by nobody. Questions about decision rights, reporting, and accountability still need to be answered.

Many board members agree that the ultimate accountability for reputation management processes needs to be located at a level of the organization whose job description is the long-term viability of the company: the board. One reason why the board is a good choice is that it can keep management's incentives for short-term solutions in check. By setting clear guidelines and emphasizing the need to safeguard reputational equity, the board can help management avoid short-sighted cost-cutting mistakes.

But the board's role is to oversee and supervise; it is not to manage the company. So, where should reputation management reside within a company's decision-making structure? A common response is that it belongs on the agenda of senior management, including the CEO. The reason that reputation management belongs on the CEO's agenda is not only that reputational risk is one of the main risks facing the company, but also that the company's reputation is one of the few sources of sustained competitive advantage. Companies with stellar reputations can charge premiums and are difficult to imitate.

One of the CEO's main tasks is to integrate reputation management into the operational processes of the business. One approach to accomplishing this task has been to create a separate corporate function: a chief reputation officer (CRO) or chief reputational risk officer (CRRO). This approach works only if the position carries weight and if the company can avoid creating yet another corporate officer with little budget and less influence. The danger in this approach is that it could create additional barriers to the integration of reputation management and business strategy, and actually hurt the process rather than help it.

An alternative is the creation of a corporate reputation council (CRC). This is a cross-functional unit composed of senior executives with actual decision-making authority. The actual composition of the council needs to mirror the organizational structure of the company. For example, a matrix organization based on global territory and product lines would have representatives from both the major territories and the business lines. In addition, the main corporate functions (marketing, finance, supply chain, HR, communication, legal, government relations, and so on) need to be represented, as reputational problems are almost always multidimensional. The decision structure must be designed to handle the complexity of such issues.

It is critical that the CRC mirrors the actual operating structure of the business. One of the reasons that Toyota was slow in responding to the 2010 sudden acceleration crisis was the lack of a truly global decision-making structure. While Toyota's economic fortunes were heavily dependent on robust US sales, decision-making was largely centered in Japan, with little input from the United States. Similarly, BP lacked a strong presence in the US regulatory and political environment, despite the fact that BP's US oil and gas assets represented more than one-quarter of the group's total annual production.

The Cost of Reputation

Good governance and decision-making structures are necessary for effective reputation management, but these alone are not sufficient. Here is why:

- Reputation consists of the perceptions of customers and other constituencies.
- In many cases, these perceptions are derived not from actual experience with the company or a deep knowledge of any given issue, but from an ever-changing mixture of opinion and information driven by the media, peer-to-peer websites, and various influencers ranging from experts to advocacy groups.
- Proactive reputation management requires companies to identify issues early, connect them with business strategy, develop prevention and preparation strategies, and implement possible changes in business practices in advance of an issue's gaining momentum.

This sequence can break down at various points. Executives may not realize the importance of reputation management for business success, governance structures may be lacking, or incentive structures may reward short-term vision. But companies may also fail to adopt effective strategies simply because they are unaware of the imminent danger. In other words, even perfectly designed governance and decision-making structures will be ineffective if they lack critical intelligence: decisions are then made in the dark.

This is the business case for investing in intelligence capabilities. Because reputation is driven by many ever-changing actors, the strategic landscape is frequently diffuse and unclear. Because successful reputational strategies need to be designed before a crisis occurs, simply surveying customers, investors, or other business partners will not do. Once customers or investors start to worry, it is too late—the deck is already stacked against the company. Therefore, in many cases, traditional business research tools such as surveys and focus groups can only measure the damage rather than prevent it in the first place. Proactive reputation management is impossible without good intelligence.

The governance structure needs to be closely connected with the intelligence function. This means that the CRC should provide strategic direction to the intelligence function and receive actionable intelligence that is directly connected to the corporate strategy. The intelligence function provides the core capabilities of issue identification, evaluation, and monitoring. The goal is both to function as an early warning system and to be able to assess the impact of corporate actions through a feedback mechanism. Without an intelligence function, the CRC will be operating in the dark and making decisions based on intuition rather than data. A company's intelligence function may range from informal monitoring of various media sources and proactive stakeholder outreach to the creation of a fully developed internal intelligence capability with its own staff and budget.

Intelligence functions are not only important for management. Given the critical role of the board as guardian of a company's reputation, it is surprising and worrisome that most corporate boards are not supported by a separate intelligence function. Such a function is ideally provided by a third party, not by company staff. Much of the critical reputational intelligence is external to the company, and it may lead board members to ask more probing questions of management.

In summary, a strategic mindset needs to be supported by effective processes. First, companies must develop a proper governance structure that should mirror the company's organizational structure. A cross-functional council is preferable to a separate corporate function unless that function is endowed with sufficient influence and resources. Second, companies need an intelligence capability. In contrast to other corporate capabilities, an intelligence system is not optional; it is essential. Reputational challenges can emerge from anywhere in the company's operations or external business environment. The lack of intelligence capabilities means that the company acts in the dark and loses its ability to manage emerging issues proactively.

The Role of People

Business leaders also need to understand that even the most advanced reputation management system is implemented by people. They need to assess the situation, evaluate risks, and then make appropriate decisions. Getting this right requires not only a strategic mindset but also values and culture in order to provide guidance to individuals. We cannot expect each employee of a company to correctly assess the reputational risk of an issue, but we can expect him or her to raise a red flag when something does not "look right."

Acting as corporate steward does not only mean doing right by customers, employees, and suppliers. It requires the ability to *think strategically*. This implies, on the one hand, viewing reputational decisions not solely as PR issues, but as decisions that are tightly connected to the company's strategy, its core competencies and values, and its distinctive position in the marketplace. On the other hand, it requires the ability to view even a familiar business decision from the point of view of people who are not specialists, but still may have strong opinions on an issue. More often than not, these opinions are not just driven by cool reason, but involve powerful emotions and passionate views of what is right or wrong behavior.

A strategic mindset also requires *situational awareness*. Reputation is essentially public. It is driven by third parties who have their own agenda. Understanding and anticipating the motivations and capabilities of these actors is essential for situational awareness. But reputational challenges are not simply the consequence of wrong decisions, accidents, or bad luck; they are frequently created by activists, interest groups, and public actors, with the goal of forcing changes in business practices through "private politics." Activists are competitors for the company's reputation. They need to be treated as seriously as competitors in the marketplace.

The last component of a strategic mindset is to avoid the *expert trap*. Becoming an expert means learning to see the world in a particular way. A doctor learns to identify symptoms and decide on a diagnosis, a poker player learns to identify "tells" of opponents that provide critical information on the strength of their hand, and a music enthusiast can pick a favorite pianist from dozens of recordings of the same piece. Acquiring and using expertise in a coordinated fashion is, of course, tremendously valuable and is at the root of the efficient organization of business processes. But, in the context of reputational challenges, it can lead us astray.

When a company collapses as a result of an earnings restatement, a trained accountant may focus on the fact that no accounting rules were violated, while everybody else will be affected by images of crying employees leaving their office for the last time. A safety engineer will point to his company's industry-leading safety standards and may be bewildered when the media focuses on one specific victim. A loan officer may view missed mortgage payments as lost revenue, while the borrower may experience them as the fear of losing the family home. The difficulty lies in the public nature of reputational challenges where company actions are evaluated by non-experts through the filter of the media. This requires decision-makers to set aside their expertise and see the situation from the point of view of laypeople in a heightened emotional state.

In summary, reputation management is not a corporate function, but a capability. It requires the right mindset integrated with the company's strategy, guided by its culture and values, and supported by carefully designed governance and intelligence processes. Developing this capability is as demanding and as challenging as developing customer focus or the ability to execute. Today's companies need to embrace this challenge.

More Info

Book:
Diermeier, Daniel. *Reputation Rules: Strategies for Building Your Company's Most Valuable Asset*. New York: McGraw-Hill, 2011.

Articles:
Minor, Dylan. "CSR as reputation insurance: Theory and evidence." Working paper. Kellogg School of Management, 2010.

Roberts, Peter W., and Grahame R. Dowling. "Corporate reputation and sustained superior financial performance." *Strategic Management Journal* 23:12 (December 2002): 1077-1093. Online at: dx.doi.org/10.1002/smj.274

Uhlmann, Eric Luis, George E. Newman, Victoria Medvec, Adam Galinsky, and Daniel Diermeier. "The sound of silence: Corporate crisis communication and its effects on consumer attitudes and behavior." Working paper. Kellogg School of Management, 2010. Online at: tinyurl.com/3n89y65 [PDF].

Notes

1. Fortune's Most Admired Companies: money.cnn.com/magazines/fortune/mostadmired
2. 2011 Edelman Trust Barometer: www.edelman.com/trust/2011/

Managing Your Reputation through Crisis: Opportunity or Threat?

by Magnus Carter

Mentor Communications, Bristol, UK

This Chapter Covers

- Definitions of risk issues and crisis.
- How risk issues can escalate into crisis.
- How to prevent escalation.
- Why some crises threaten your reputation more than others.
- How to safeguard your reputation in a crisis.
- What to say and what to do.
- The important role of the media, including social media.

Introduction

We know that no organization is immune from crisis. The one thing that is predictable about crises is that they *will* happen, and a good crisis management plan is essential best practice. Yet my experience as a crisis communications consultant tells me that many organizations, including some substantial ones, do not have such a plan. And of those that do have a plan, surprisingly many consider only the direct threat of disaster, catastrophe, and physical emergency, and how to prevent such events from interrupting business.

My argument in this chapter has the following three fundamentals.

- That the greatest threat to organizations in crisis is often the damage caused to their reputation. You may be able to recover your ability to deliver product, for example, but if no one trusts your product any longer, the exercise is pointless.
- That reputation is most adversely affected where crises arise from risk issues that might have been foreseen and better managed. However, even disasters in which you may initially be seen as the "victim" have a nasty habit of throwing up challenges to your reputation, depending on how you manage the event.
- Managing your organization's reputation needs to be at the center of crisis management and recovery. It is not something you can "bolt on." Being seen and heard to manage risk issues and crises well can enhance your reputation and limit or mitigate damage from the event. So dealing with the media, including online sources, is not optional—it is essential.

Crisis and Reputation

The key to understanding how much of a threat to your reputation a crisis is likely to create and where that threat is likely to arise is to consider the following question: What is within your control, and what is beyond it?

Figure 1 illustrates the point. It was devised by Samuel Passow, a Harvard-educated journalist, author, and accredited mediator who is the head of consultancy services and executive training at the Conflict Analysis Research Centre at the University of Kent. The types of crisis that appear toward the left-hand end of the spectrum are the most likely to dent your reputation. In those toward the right, the early assumption is likely to be that you are the "victim." It is vital that crisis management plans do not focus entirely on the right-hand end of the spectrum. Always allow for mismanagement. And do not listen to the argument that such and such will never happen.

Figure 1. The spectrum of crises. (Devised by Samuel Passow)

The spectrum of crises

Within your control: Management failure; Ownership battles

Influence but not control: Confrontation with activists; Industrial accidents; Product danger Health and safety

Beyond your control: Civil disorder; Natural disaster; Crime and terrorism; War

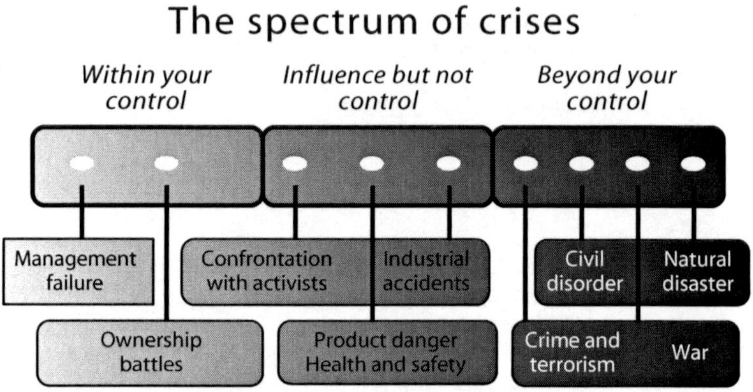

Of course, how you manage a crisis is always under your control, so a crisis that begins at the right-hand end of the spectrum can move to the left if you are perceived to be mismanaging the situation, or if it is suspected that better management might have prevented or mitigated the crisis.

For example, in the Japanese earthquake and tsunami of 2011, sympathy for the plight of the nation and those affected was universal, as was admiration for the stoicism and robustness of individual and official responses. However, that did not prevent important questions being raised about Japan's nuclear safety policy in the light of what happened at Fukushima—questions that had serious repercussions for the reputation of the nuclear industry worldwide, as shown by Germany's almost instant decision to announce the ending of its nuclear energy program.

The Fukushima Daiichi plant was badly damaged by the 9.0 earthquake that struck on March 11, 2011. The Tokyo Electric Power Company, which owns the plant, had previously carried out geologic and sonic surveys to assess the power station's resistance to such events. They had put in place a number of additional precautions as a result of learning from a fire at another plant in the northwest of the country when it was hit by a magnitude 6.6 earthquake in 2007.

But Japan's Nuclear and Industrial Safety Agency said that, despite the surveys, it appeared that officials at Fukushima had not considered the scenario that a tsunami might hit the power plant at a time when they would need to use the diesel backup

Managing Your Reputation through Crisis: Opportunity or Threat?

generators intended to provide emergency power to the reactor cooling systems. Fuel tanks for the generators, positioned at ground level just yards from the seafront, were among the first parts of the facility to be destroyed by the huge tsunami wave that swept inland following the earthquake.

Dr John H. Large, a UK-based independent nuclear engineer and nuclear safety expert, told the *Daily Telegraph*: "These plants should be designed to be resistant to tsunamis, but it appears they did not consider that a tsunami would hit the plant when they were using the back-up generators. The buildings will have been built to withstand a tsunami, but it appears the back-up generators were not." (Gray and Fitzpatrick, 2011).

This highlights an important aspect of crisis planning: the law of unfortunate coincidence, which says that whenever one thing goes wrong, so will another.

Risk Issues and Crisis

Mike Seymour and Simon Moore, in their work *Effective Crisis Management*, classify crises as either cobras or pythons. Cobras strike suddenly and take you by surprise, and are more likely to appear at the right-hand end of our spectrum. Pythons creep up over time and slowly crush you, and are more likely to appear at the left-hand end of the spectrum. Crucially, pythons do not start out as crises, and there is time to spot them and deal with them before strangulation occurs: at this stage, they can be classified as *risk issues*. But how to spot them before they become *crises*?

Let's start by defining what are in fact two stages of the same continuum. The *Journal of Management* Studies offers the following definitions:

- *Risk issue*—A point of conflict between an organization and one or more of its audiences, or a gap between corporate practice and stakeholder expectations. (In other words, a risk issue arises when you don't do what you say you do, or what others have a right to expect you to do).
- *Crisis*—An organizationally-based disaster which causes extensive damage or disruption and involves multiple stakeholders.

Every time we fail to meet customer expectations, we have raised a risk issue. Managed well, there is no lasting damage. Managed badly, the risk issue begins to escalate: other customers begin to join in, perhaps the media take an interest, and the risk issue tips over into crisis.

A notable and historic instance of this process is provided by the experience of Shell UK in the Brent oilfield in the North Sea. Brent Spar was a crude oil-storage and tanker-loading buoy for 15 years until it was decommissioned in 1991, having reached the end of its useful life.

An environmental outcry ensued when it emerged that Shell UK Exploration and Production (Shell Expro) was planning to dispose of the structure by sinking it in the Atlantic Ocean. As well as Greenpeace's occupation of the rig off the Scottish coast, many people boycotted Shell products in a campaign which the company said lost it millions of dollars.

Risk Management in an Uncertain World

The publicity led the oil company to drop its plans in 1995, and no oil structures have been dumped at sea since then. In July 1999 European nations agreed to ban the dumping of offshore steel oil rigs.

Although Shell had carried out an environmental impact assessment in full accordance with existing legislation and firmly believed that its actions were in the best interests of the environment, it had severely underestimated the strength of public opinion. As a result, the oil industry now faces an expensive and environmentally questionable requirement to return all North Sea oil installations to shore when redundant, in contradiction of most scientific evidence.

Shell was particularly criticized for having thought of this as a "UK" (or even "Scottish") problem and for neglecting to think of the impact it would have on the company's image in the wider world. To quote the *Financial Times*: "In hindsight, Shell failed to detect the extent of public concern in continental Europe or to win adequate support for its argument that the best place for the Brent Spar was in a deep trench in the Atlantic. As a result, years of careful cultivation by Shell of an environmentally friendly image have been thrown away." And here's the *Wall Street Journal*: "Shell made a strategic error. In a world of sound bites, one image was left with viewers: a huge multinational oil company was mustering all its might to bully what was portrayed as a brave but determined band."

So, the issue of how best to dispose of a redundant piece of equipment in the most environmentally sound and safe way escalated into a crisis of public confidence—a threat to the company's reputation. The question is, how does this happen?

Crisis Escalation

In many cases, issues develop as a result of external processes or legislation, management directives, or working practices. Often, legislation or guidelines are simply viewed as "something that has to be done"—and although the operational implications are considered, the cultural and social impact is missed or ignored. This is a key element in the process known as "crisis incubation." The incubation of issues until they become crises results from a number of cultural factors.

- *Management values and perceptions:* The primary beliefs and values of an organization or its management may restrict decision-making or planning. In particular, autocratic management styles at the top tend to permeate through organizations and lead to contrary opinions and information being withheld. Processes and controls will only be created and implemented for perceived issues that fit the organization's demonstrated values (which are often not in tune with those which are publicly declared).
- *Denial of possibilities:* Organizations and managers will often not look beyond the identified issues and will deny that the organization requires processes to handle issues that may actually occur. This leads to…
- *Internalization:* Managers ignore or refuse to accept the opinions of people outside their own organization. This may be because they do not believe that outsiders have the expertise, or because outsiders' opinions or values do not match their own, or because they do not want their own expertise questioned.

Managing Your Reputation through Crisis: Opportunity or Threat?

- *Disempowered employees:* The combination of autocratic management and a failure to acknowledge external perspectives will usually result in a workforce that is reluctant to pass on knowledge gained on the shop floor or in encounters with customers. In this way, important information fails to reach decision-making levels, important decisions are missed, and a downward spiral of employee disenfranchisement begins.
- *Compliance:* Regulations or processes are, or are perceived to be, outdated or inappropriate for the particular organization, individual, or situation. As a result, the rules are not followed.
- *Monitoring:* Issues and processes are implemented but not monitored or evaluated to judge success, measure against original objectives, or assess impact or the need to change. Without constant monitoring, it cannot be established whether issues have been resolved or, at least, kept under control, or are actually heading toward crisis.
- *Blame culture:* Blame and finger-pointing hinders learning and can cause further unnecessary damage to the organization.

In the case of Shell and the Brent Spar, it is likely that the first three of these factors were at work. Shell knew it was right, but failed to recognize public perception and to engage in debate to put across its argument until it was too late and the company was forced into an embarrassing climbdown.

Avoiding a Crisis Culture

Arguably, a crisis culture stems from an organization's inability to address, identify, or resolve issues. An organization with a culture which is open to internal and external debate and advice about issues as they arise, which is prepared to initiate change as a result of that advice or debate, and which is prepared to communicate clearly how issues have been resolved is an organization that is far more likely to avoid crises in the first place, or to be able to defuse them quickly when they occur.

One of the most significant factors in both crisis incubation and postcrisis learning is blame. Identification and resolution of issues is far more difficult in a culture that fosters blame. Whistleblowers only exist—and are seen in an entirely negative light—because organizations lack transparency and fail to listen to, or address, issues raised in the day-to-day working environment by people at all levels within them. "Blowing the whistle" is an act of last resort by people who can no longer watch bad or dangerous practices being conducted by their colleagues. Whistleblowing marks the point at which an issue has incubated into a crisis, and a no-blame culture will help to avoid this.

The identification of issues—which may often be associated with someone's weakness or failure—is hindered by the fear of finger-pointing, blame, and accusation. In a no-blame culture, however, your ability to make rapid, clear decisions will be enhanced by a reduced fear of "getting it wrong."

In a crisis, the media and other external parties will be looking for someone to blame. *Don't give them the opportunity.*

Risk Management in an Uncertain World

How to Avoid a Blame Culture

- Focus on the culture or processes that have led to the issues, rather than looking at one individual. Remember, 70% of crises are thought to arise from organizational rather than individual issues.
- Accept and encourage constructive suggestion and criticism from all levels of the organization, and even from external stakeholders. Staff or stakeholders at the "sharp end" can often identify issues far more swiftly and clearly than middle or senior managers.
- Listen—don't dismiss advice, comments, suggestions, or criticism. Listen to what is being said. In particular, if the comment is aimed at an individual, look beyond the person and seek out the underlying issue.
- Focus on the bigger picture—many issues continue to incubate because organizations deal with the small, easily resolved technical issues and ignore the larger, more complex and difficult cultural issues.
- Ensure the involvement of representatives of all levels of the workforce when developing, enhancing, and reviewing processes or key operational activities.
- Provide opportunities for formal discussion in a nonthreatening environment—hold team meetings away from the office, or in a relaxed area. Set ground rules that encourage people to talk freely about issues rather than personalities.
- Provide opportunities for informal, ad hoc contact—in a kitchen or relaxation area, for example. Don't dismiss issues that are raised in informal discussions, as these are often the most revealing conversations. But understand the difference between informed comment and gossip.

Managing Issues to Avoid Crisis

The discipline of identifying and managing external risk issues has grown in importance. It received a kick-start in the 1960s from two US sources. The first was Rachel Carson's groundbreaking book *Silent Spring*, which argued that uncontrolled use of pesticides such as DDT was harming and killing not only animals and birds, but also humans. The second was Ralph Nader's *Unsafe at Any Speed*, a book which claimed that many American automobiles were a danger to drivers and pedestrians alike. The first marked the start of environmentalism, and the second, of consumerism—both of them forces that organizations would ignore at their peril.

Since then, we have witnessed the rise of consumer litigation and a growing recognition on the part of private business that reputation is its most important asset. As discussed elsewhere in the present volume, this becomes ever more important and urgent in the digital age, when news travels so fast and where consumers and activists can have influence and access on a par with companies and governmental agencies. I would also argue that social media have brought about a loss of deference toward organizations and an increasing emphasis on the "me" culture.

More and more organizations are rising to these challenges by putting reputation management at the center of management processes. This is sometimes given focus by establishing a reputation management group, drawn from all disciplines and all areas

Managing Your Reputation through Crisis: Opportunity or Threat?

of the business, whose job it is to identify the issues that create risk for the organization. This should not be seen simply as a defensive operation: it enables your organization to support and adapt strategies, plans, and operations and can enhance your abilities to capitalize on opportunities, reduce risk, and secure competitive advantage.

Importantly, a reputation management group must not be allowed to degenerate into a talking shop. Neither can its task be delegated to the corporate communications function. To succeed, it must be locked into the decision-making process, and it must have the support and input of all management disciplines, recognizing that managing reputation is a fundamental task for all managers. Key features of a reputation management group are set out in the next section.

The Reputation Management Group

Members of the group should:

- be drawn from all disciplines/areas of the business;
- be able to draw on external expertise/advice if required;
- have decision-making powers, or board access.

The functions of the group are:

- to monitor public policy/legislation;
- to monitor key external issues;
- to monitor trends in public opinion.

The goals of the group are:

- to understand potential impacts of the above three items on the business;
- to adjust strategy, plans, and processes accordingly;
- to develop and communicate organizational positions/responses, internally and externally.

The multidisciplinary approach to risk management represented by a properly constituted reputation management group is essential. Without that, definitions of risk tend to become too narrow, often focusing on health and safety issues, for instance.

This narrowness of definition was surely one factor behind the problems that beset UK banks in 2008–09. Few organizations take risk management more seriously, and certainly none has more employees with the words "risk manager" somewhere in their title—and yet, if the Financial Services Secretary to the UK Treasury Paul Myners is right, on Friday, October 10, 2008, the country was "very close" to a complete banking collapse after major depositors attempted to withdraw their money *en masse*. The focus of banks' risk management has developed and evolved over the past few years. That focus is now increasingly on reputational, regulatory, operational, and strategic risk, as well as the more traditional credit and market dimensions of risk. In 2008 that focus was still too narrow, and a blind eye was being turned to some of the banks' more aggressive lending.

Risk Management in an Uncertain World

On the other hand, perhaps the banking crisis simply illustrates a more general point. While identifying and managing risk issues can bring real reputational benefit and stave off many potential crises, there remains the problem that was identified by British Prime Minister Harold Macmillan toward the end of his premiership in 1963. Asked what he thought most likely to blow a government off course, he replied: "Events, dear boy, events." There will always be crises that come out of the blue. Indeed, we are all fallible, and therefore there will always be risk issues that no one spots until it is too late.

What Issues?

The issues that are relevant or threatening will vary not only from organization to organization, but also with time and place. Here are just a few you may need to consider.

- Climate change: flood risks, carbon footprint, carbon offset, etc.
- Credit and interest rates
- Health and safety (compliance)
- Corporate governance
- Ethical standards
- Terrorism
- Crime and security
- Disaffected youth
- Flu or other epidemics
- Transport costs, overcrowded roads, etc.
- Energy prices
- House prices (impact on key workers)
- Education standards
- Board pay
- Loss of key staff
- Outsourcing
- Call centers and customer service

How to Recognize Your Crisis

When crises are cataclysmic events ("cobras") they are comparatively easy to identify. If your factory burns down, that is clearly a crisis. Most crises, including all those that arise from risk issues, are less easy to define as such: there is more usually a continuum from controlled issues to out-of-control crises ("pythons").

The section below "How serious?" offers a checklist for evaluating your crisis and deciding a proportionate response. There is perhaps a simpler way of spotting the tipping point between issue and crisis: this is when you receive more than one media call, or when online discussions start to attract widespread responses. By this time, you need to be acting and communicating.

How Serious?

The issues that are relevant or threatening will vary not only from organization to organization, but also with time and place. Here are just a few you may need to consider.

Managing Your Reputation through Crisis: Opportunity or Threat?

1. *Risk:* How much is at risk? Where are the risks?
2. *Control:* To what extent is the organization responsible for the crisis? Who else is responsible?
3. *Affected parties*: Who is affected by the crisis? More than one stakeholder group? How?
4. *Trajectory:* Is the crisis likely to escalate?
5. *Time:* How much time is there to maneuver?
6. *Interest:* Is the crisis likely to foster outside attention beyond the parties directly affected? Are the media likely to take an interest?
7. *Spillover:* Is normal business operation interrupted or affected? Are suppliers/customers likely to lose confidence? Will there be a bottom-line effect?
8. Scope: How many choices do we have? What is the quality of those choices?
9. *Options:* What is the organization's ability to influence the situation? Which parties in the situation need to be influenced?
10. *Communication:* What channels of communication will best help you to reach the audiences identified in indicators 3 and 9 above?

Communicating Your Crisis to Safeguard Reputation

Communication is not an optional extra in a crisis. It needs to be part of the management task, and it should be consistent with the actions you are taking. What you say must reflect what you do—and you won't be doing nothing, so you should never be *saying* nothing.

The temptation is to keep quiet in the early stages of a crisis, because you are likely to know very little: "We don't know enough about what's happening, we don't have solid information, so we can't really say anything." However, there are things you can say when you know very little, things that will help to manage your reputation from the start. And it is vital to say these things at the earliest possible opportunity. As long ago as 1970 the historian and satirist C. Northcote Parkinson wrote, "The vacuum created by a failure to communicate will quickly be filled with rumor, misrepresentation, drivel and poison." In an age of 24-hour news and globally accessible social media, that is truer than ever.

When it comes to giving information, there should be a presumption of openness. The question should be "Is there any good reason why we shouldn't reveal this?" rather than the more usual "Do we really need to tell them this?" A general spirit of openness and honesty will always enhance your reputation. Any suspicion that facts are being hidden will do the opposite.

Until the advent of the blogosphere and Twitter, the traditional media represented the only universal way of communicating with all your stakeholders at once. For the time being, it is still the case that traditional media coverage (TV, radio, press) is more influential than social media coverage when it comes to effects on your reputation.

As discussed elsewhere in this book, this may not remain true for ever, and it may already be untrue in some sectors (e.g. online retailing, and perhaps consumer electronics). It is important to understand also the close relationship between social media and journalism: most newsrooms now regularly monitor Twitter feeds and all are involved in and interact with blogs and discussion forums.

Risk Management in an Uncertain World

Organizations need to use all available communication channels to engage with stakeholders. In the case of both traditional and social media, conversations are more effective where there is an existing relationship. Twitter followers will not trust you if you simply create an account to answer criticism that is already out there. Journalists will have no reason to trust you if they have never heard from you before.

What to Say?

Whatever channels of communication you are using, what you choose to say is crucial. The good news is that there is a magic formula to help you to decide what to say. It can be applied to any type or scale of crisis, and indeed the same formula works well when speaking about risk issues, for example customer complaints. Applied wisely, and supported by real action, this formula can be a powerful aid in preventing issues from escalating into crises.

The formula is encapsulated in the acronym CARE. Whatever the crisis, your media statements, interviews, blog postings, etc.—especially early in the crisis management process—should always be based on the following.

» *Concern:* For example, understanding the point of view of protesters, or showing sympathy for families of the bereaved. Do not be afraid of the word "sorry" if it is appropriate. However, note that "We are sorry this has happened" is not the same message as "We are sorry we got it wrong," but in either case the word has the power to defuse volatile reactions. The UK National Health Service Litigation Authority recognizes this in its guidance to hospital trusts, advising that, where it is clear that there has been a medical mistake, hospitals should apologize to those affected and their families at the earliest opportunity.
» *Action:* Your audience needs to know that you will be doing something, and as much as possible about what that might be. At the very least it should be told that there is to be some sort of investigation. Good crisis planning, to some extent at least, enables you to know what happens next.
» *REassurance:* This means, for example, saying that lessons will be learned, or that you have a good safety/security record, or that you have contingency plans in place. This message must be accompanied by the preceding two messages if it is not to come across as defensive and suspicious.

Who Should Say It?

Once one has accepted that there is positive advantage to be gained from communicating in a crisis, this question answers itself. Clearly, the best person to represent an organization when it wants to be taken seriously, or when it wants to indicate how seriously it is taking a situation, is the person at the top—probably the chief executive or chairman.

There are some exceptions to this rule, when you may decide to engage at a lower level. You may want to deliberately signal that the perceived crisis is not as serious as others may think. You may want to speak from specific expertise: for example, the technical director may be more appropriate in certain situations. Or there may be practical difficulties: the chief executive is on the other side of the world. Finally, you may believe that your chief executive does not possess the skills or aptitude needed to represent you well on camera. This is a matter of judgment, but the presumption

Managing Your Reputation through Crisis: Opportunity or Threat?

should be that the most senior person available should be your public face, and that you should avoid the anonymity of "a spokesperson." This applies equally to issuing statements to the media or to content on your website or blog.

It is worth saying that no one is a "natural" at this. Getting the right combination of content, tone, and style of delivery is essential, and that takes practice. Media training is therefore a prerequisite, as is regular exercise testing of your crisis communications plan.

There were many factors that made life difficult for BP in the aftermath of the 2010 incident in the Gulf of Mexico, when the Deepwater Horizon rig blew up, killing 11 men and causing an oil spill that created massive pollution for more than three months. BP seemed to do the right thing in putting forward its then chief executive, Tony Hayward. The problem was that he was also in direct overall charge of BP's recovery operation. The lesson here is to keep the role of head of crisis recovery completely separate from that of spokesman.

In short, your public face should be the most senior person who is capable of presenting the best possible "face" of the organization and who is not operationally locked into the process of crisis management.

Case Study

Classic Case Study
The Kegworth Air Crash

Serious air crashes tend to lead to a dip in ticket sales, sometimes related to a specific type of aircraft, sometimes to the airline involved, and sometimes both. For instance, the loss of an Indonesian budget airline Adam Air Boeing 737 on New Year's Day, 2007, in which 102 people died, led to a loss of more than 30% of the airline's sales, which took several months to recover.

However, in the case of the January 1989 crash of a British Midland Boeing 737 on to a superhighway embankment near East Midlands Airport, ticket sales actually increased in the following weeks—even though the crash killed 47 people. The communication around this incident by British Midland provided a model for all future crisis handling.

The airline chief executive, Michael Bishop, began giving interviews within an hour of the incident, when little was known about the circumstances. In the absence of facts, Bishop focused on expressing how he felt about what had happened and what he was going to do about the situation—in other words, he began to manage the flow and content of news to the media. His content was in line with the CARE formula. As a result, the airline was perceived to be caring and responsible—it was seen as part of the solution, rather than part of the problem.

There was little information in what Michael Bishop said, but he followed the golden rules:

- begin communicating at once—take the initiative;
- communicate from the top of the organization;
- treat all media seriously and don't neglect any outlet—especially those in your sphere of operation, including the regional press where you are located.

Risk Management in an Uncertain World

Summary and Further Steps

- When it comes to risk issues and crisis, *perception is reality*. Know and act on what people *think*.
- Risk issues need to be actively identified and managed to protect and enhance your reputation.
- A multidisciplinary reputation management group, with decision-making powers, can provide a focus for this.
- Well-managed risk issues reduce the likelihood and impact of crises.
- Foster a no-blame culture to ensure that risk issues are flagged up and the resulting policy is respected internally.
- Despite all your best efforts, crises will happen.
- Communicate quickly and openly in a crisis, using the CARE formula, and through all available channels. Aim to be seen as part of the solution, rather than the problem.
- Exercise regularly to test and revise crisis communication plans and to identify and support your spokespeople.

More Info

Books:

Anthonissen, Peter F. (ed). *Crisis Communication: Practical PR Strategies for Reputation Management and Company Survival*. London, UK: Kogan Page, 2008.

Carson, Rachel. *Silent Spring*. Boston MA: Houghton Mifflin, 1962.

Coombs, W. Timothy. *Ongoing Crisis Communication: Planning, Managing, and Responding*. 3rd ed. Thousand Oaks, CA: Sage Publications, 2011.

Fearn-Banks, Kathleen. *Crisis Communications: A Casebook Approach*. 4th ed. New York: Routledge, 2010.

Griffin, Andrew. *New Strategies for Reputation Management: Gaining Control of Issues, Crises & Corporate Social Responsibility*. London, UK: Kogan Page, 2009.

Holmes, Anthony. *Managing Through Turbulent Times: The 7 Rules of Crisis Management*. Petersfield, UK: Harriman House, 2009.

Nader, Ralph. *Unsafe at Any Speed: The Designed-In Dangers of the American Automobile*. New York: Grossman, 1965.

Parkinson, C. Northcote. *The Law of Delay: Interviews and Outerviews*. Boston, MA: Houghton Mifflin, 1971.

Regester, Michael, and Judy Larkin. *Risk Issues and Crisis Management in Public Relations: A Casebook of Best Practice*. 4th ed. London, UK: Kogan Page, 2008.

Seymour, Mike, and Simon Moore. *Effective Crisis Management: Worldwide Principles and Practice*. London, UK: Cassell, 2000.

Smith, Denis, and Dominic Elliott (eds). *Key Readings in Crisis Management: Systems and Structures for Prevention and Recovery*. New York: Routledge, 2006.

Ulmer, Robert R., Timothy L. Sellnow, and Matthew W. Seeger. *Effective Crisis Communication: Moving from Crisis to Opportunity*. 2nd ed. Thousand Oaks, CA: Sage Publications, 2011.

Managing Your Reputation through Crisis: Opportunity or Threat?

Article:
Gray, Richard, and Michael Fitzpatrick. "Japan nuclear crisis: Tsunami study showed Fukushima plant was at risk." *Daily Telegraph* (March 19, 2011).
Online at: tinyurl.com/43f27ue

Report:
Shell. "Brent Spar dossier." 1999. Online at: tinyurl.com/3mvaka7 [PDF].

Websites:
Greenpeace on Brent Spar: tinyurl.com/3hm9wz2

Institute of Risk Management: www.theirm.org

PR Media Blog on crisis communication: tinyurl.com/3zmruq9

Crisis Management and Strategies for Dealing with Crisis
by Jon White
Consultant, London, UK

This Chapter Covers

- Crises and their characteristics—what distinguishes crises from disasters, emergencies, and other exceptional situations.
- Crisis management—phases in crisis management, and the requirements in each phase for effective management.
- Techniques for anticipating crises, and preparations for dealing with them.
- The benefits to be gained by routine management from paying attention to crisis management.
- The demands made on management at times of crisis and the importance of psychological preparation.
- Practical conclusions.

Introduction: Crises and Their Characteristics

The term "crisis" is much used in the media coverage of events, in public discussion, and by organization leaders describing the situations that they face and try to manage. The consequences of failure to manage these situations may prove more or less damaging to their organizations. However, not all situations that are described as crises should be labeled as such, and it is important to distinguish real crises from other situations as a first step toward managing them as effectively as possible.

Before looking at the characteristics of crisis situations, we will examine a situation—the Deepwater Horizon oil spill in the Gulf of Mexico—that was recognized at the time as a crisis for the organization involved to draw out the features that define it as a crisis.

Case Study

BP in the Gulf of Mexico, 2010

On April 20, 2010, an oil rig—the Deepwater Horizon, which was operated for BP in the Gulf of Mexico by Transocean—was rocked by an explosion which killed 11 workers and allowed oil to escape from the wellhead, 40 miles off the coast of Louisiana and 5,000 feet below the surface of the sea. Oil flowed into the water for 87 days before the well was finally capped.

After the event, BP recognized that it was being subjected to unprecedented media and political scrutiny. For President Barack Obama the stakes were high—his predecessor, President George W. Bush, had been heavily criticized for his perceived slowness of response to the impact of Hurricane Katrina on the same US state in 2005.

In addition to the deaths of employees, the accident caused damage to the environment and economic damage to the affected states. For BP, the accident was a corporate crisis that threatened the existence of the company, wiping 50% off its market value and forcing it to set aside funds to meet potential claims against the company.

The management response to the accident and its aftermath was a reaction to a complex interplay of issues:

- the technical problems associated with attempts to cap a well in a deepwater operation;
- the organizational requirements of dealing with several partner and supplier organizations that were not part of BP, but were providing services, such as the operation of the rig, to BP;
- the number of involved stakeholders, from the US federal government, to local governments and the communities affected.

BP's CEO, Tony Hayward, became personally involved in the company's response to the accident, suggesting that he would not leave the site until the problem had been solved. He was, however, to be heavily criticized in his role and for a number of statements that he made to the media during the course of managing the company's response to the accident and its aftermath. His credibility was damaged, and he was replaced as CEO of the company as it sought to retrieve its reputation after the well was finally capped.

Media scrutiny of the company was intense, with attempts to deal with the problem of the oil leaking into the sea and to explain the company's actions under constant surveillance by the press and broadcast media, and in discussions that took place on the social media.

BP's experience in dealing with the Gulf of Mexico incident and its aftermath is already a classic case study, taking its place alongside other oil industry case studies dealing with Shell's experience with the Brent Spar in the mid-1990s and Exxon's disastrous Alaskan oil spill from the company's vessel, the Exxon Valdez, in 1989.

But what made the accident a crisis for the company? We will consider this question in the next part of the chapter.

Distinguishing Crises

In the academic literature on crisis management, crises are distinguished from other serious situations faced by organizations by a number of defining features.

Level of Threat

First, a crisis has to be serious in its consequences, involving loss of life, or threat to life, and serious injury or the possibility of serious injury, to numbers of people. It will also involve actual or possible damage to property. The first defining feature of crisis is the *level of threat* involved. The threat may also extend to the organization's existence, vital interests, or "raison d'être."

How an organization responds to a crisis will have an impact on its reputation, which is an aggregation of perceptions of its performance. Failure to perform to expectations

Crisis Management and Strategies for Dealing with Crisis

at a time of crisis will result in damage to a company's reputation, which in some cases will be terminal.

There are also well-documented examples of corporate performance at times of crisis that enhance an organization's reputation. An example is the performance of Swissair (the airline went bankrupt in 2002) at the time of the crash of one of its aircraft off Nova Scotia, Canada, in 1998, a crash in which all 229 people on board the aircraft died. The airline's competent and sensitive handling of its obligations to those affected by the crash protected, maintained, and enhanced the airline's reputation.

Time Pressure
A second defining feature of crisis is *time pressure*. Threat or damage must be dealt with as soon as possible to minimize their consequences—whether these are further loss of life, threat to life, or damage to property. A good example here is the Japanese response to continuing threats of radiation leakage from the Fukushima nuclear plant damaged in the tsunami which followed an earthquake off Japan in March 2011.

Stress
A third defining feature of crisis is *stress*, as experienced by the decision-makers who have to deal with the situation. In crisis, their response will be marked by surprise, uncertainty, and flawed decision-making. In many crisis situations, the observation will be made that decision-makers are "out of position"—not where they need to be to deal with the situation as it develops.

Crises are unexpected, unforeseen events, which distinguishes them from disasters. Disasters, although serious in their consequences and sharing characteristics with crises, can be anticipated and plans made for dealing with them. For example, countries that experience earthquakes can prepare to respond to them when they occur, and take steps to minimize their consequences, by setting up early warning systems and buildings that can withstand earthquake damage.

Crisis management depends on classifying events or situations that will constitute crises for any organization. They will pose threats to life and property, and to the existence and reputation of the organization. They will need to be dealt with quickly to minimize their consequences, and they will make difficult demands on decision-makers and others who have to deal with them.

The Deepwater Horizon Incident as an Example of a Crisis
The BP case shows all these characteristics—the accident killed numbers of employees, damaged equipment, and created a huge oil spill that impacted the environment, the communities around the Gulf, and the livelihoods of their inhabitants. The company's vital interests—its reputation and ability to operate in deepwater environments in the United States and worldwide—were called into question, as were the abilities of the managers who tried to deal with the accident and the oil discharge and their consequences.

In addition, the crisis was played out in public, against sustained questioning and criticism from government, the media, the public, and investors in the company.

Risk Management in an Uncertain World

Categories of Corporate Crisis
The Deepwater Horizon incident was a "perfect storm" of a crisis. At end of the film *The Perfect Storm*, the fishing boat that is trying to ride through the storm is overwhelmed by the sea. How can organizations save themselves from being overwhelmed by such "storms"? They need first of all to understand what they may be up against, by trying to understand what, for them, will constitute crises so that they can prepare for them.

Lerbinger (1997) has outlined a number of broad categories of corporate crises.

- *Crises involving technology:* In a world that is increasingly dependent on technology, when technology fails the consequences may be catastrophic. Examples of these crises are provided by the difficulties at the Fukushima nuclear plant in 2011 following the earthquake and tsunami in Japan, and the Bhopal industrial accident in 1984 when gas released from a Union Carbide plant in Bhopal, India, caused many fatalities.
- *Crises arising from confrontation:* These are caused when groups confront corporations or other authorities and criticize their actions, or go to more extreme lengths to express their opposition or discontent. An example of this is provided by the action of animal rights pressure groups in mounting strong, or even violent, opposition to companies that make use of animals in product testing or research.
- *Crises caused by malevolence:* These are crises caused by the malevolent actions of individuals or groups, such as terrorist groups placing bombs in unlikely locations to bring about maximum disruption of business and everyday life. The 7/7 attacks on London's transportation systems—the subway and buses on 7 July, 2005—are an example of a crisis created by malevolence.
- *Crises of management failure:* These are crises caused by management groups within the organization failing to carry out their responsibilities. Arguments have been made that the so-called financial crisis of recent years could be attributed to failure on the part of managers of financial institutions to manage the risks that their organizations were taking.
- *Crises involving other threats to the organization:* Examples include unexpected takeover bids.

Crisis Management
How can the unexpected be prepared for and managed? How can the manager plan for that which cannot be planned for? How much attention should be given to planning for events or situations which may never happen? These are practical and difficult questions for managers thinking about crisis management to address.

Crisis management can be approached in terms of three phases of crisis management, and the requirements for managing each phase can be explored. Crisis management involves:

- crisis planning;
- crisis management—management of the crisis when it occurs;
- managing the aftermath of a crisis.

Crisis Management and Strategies for Dealing with Crisis

Crisis Planning

With hindsight, it is easy to ask whether or not an organization in crisis did enough to plan for the situation that arose. Questions along these lines were asked about BP's preparations for the oil spill which occurred in the Gulf of Mexico.

Crisis planning is an exercise in thinking the unthinkable—what, given the organization's interests, operations, plans, and links to groups such as suppliers, the community, and interest groups, could possibly go wrong for the organization? This exercise is one that should involve the most senior managers of the organization, and it may be difficult to get their attention and time for this—some will feel that they need not concern themselves with events that are unlikely to happen.

The exercise will typically generate a list of between 20 or 30 events and situations, which can be categorized and judged in terms of their likelihood of occurrence and impact. They may be placed on a scale of severity, ranging from crises (events or situations that affect the central interests or existence of the organization, involve threats to life and property, and require prompt action), disasters (similar, but more predictable), emergencies (less threatening situations, predictable, but still requiring prompt action), and exceptions to routine operations (predictable, requiring prompt remedial action). When events and situations have been categorized, plans can be made to deal with each category.

Crises require specific plans, which set out the situations that the organization has identified as crises. The plans set out arrangements for mobilizing the organization—identifying its senior managers and others who will be involved in managing the crisis situation. Crisis management requires two management teams—one to make decisions regarding the organization's response to the crisis, the other to implement the decisions made by this group and ensure that the organization deals with the demands of the situation. Arrangements have to be made for these two teams to work alongside each other, from facilities suitable for their work and with sufficient communication links between the teams, to the rest of the organization and to the outside world.

Crisis plans will identify roles in crisis management and the individuals who are to carry them out, and will set out arrangements for filling these roles over periods of time.

Communication at a time of crisis needs special attention. Will the communication links required be in place and sufficient? Who are the groups of people with whom the organization will need to communicate? What can and should be said at the outset in any crisis situation to groups such as the affected stakeholder groups (for example, the relatives of any employees injured in a crisis situation), or the media?

Some of this material can and should be prepared before any crisis situation develops. This can be incorporated into website content, into specific websites prepared for opening to the public at time of crisis, or into sections of already accessible websites. A recent development in crisis communication has been to take into account the part played by the social media, allowing for stakeholder and public comment on the crisis situation as it develops. Planning for crisis communication now requires planning for possible developments in the discussion of crisis situations on such media, where the organization needs to decide what its approach to these channels will be before the crisis situation develops.

Once developed, crisis plans have to be tested, rehearsed through simulations, and kept up to date. It is particularly important to keep up to date the information about individuals who are nominated to play specific roles. All details of the plan can be tested against experience from simulations, where situations similar to those foreseen in the planning phase are presented to the organization to test the response of crisis management teams and the organization itself.

Managing the Crisis Itself

Despite the importance of planning, during a time of crisis the plans that have been developed may not fit the actual crisis situation. The situation may differ from any that was considered in the planning phase, or the crisis may overturn key features of existing plans. For example, the plans may call for the use of specific facilities from which the crisis is to be managed, but the actual crisis may involve the destruction of facilities intended for this use. Back-up plans should be in place, but there may be a need for improvisation at a time when the quality of decision-making is degraded by time pressure and stress.

Crisis management under these conditions is improved by preparation—although managers do not have specific preparation, they are prepared for stress and the difficulties of decision-making under stressful conditions. A characteristic of decision-making under these conditions is "group think." This is a term developed by the psychologist Irving Janis in the early 1970s to explain faulty group decision-making. In this, groups rely on their own resources and focus on group interests rather than on information from outside that might contradict prevailing views within the group. Group think can be avoided by ensuring an adequate flow of information to decision-making groups, which in turn requires sufficient communication links for these groups.

Organizations that face the possibility of demanding crisis situations need, where resources allow, to have in place advisors who at a time of crisis can monitor, comment on, and raise questions about the quality of decision-making and the need for additional information to inform decisions.

Managing the Aftermath of Crisis

After the immediate crisis situation has been brought under control—for example, after BP finally capped the well in the Gulf of Mexico—organizations have to manage the aftermath. This may involve correcting practices that led to the crisis, compensating those who have been affected, and reestablishing external confidence in the organization. Managing the aftermath requires longer-term effort, and will also involve communicating with important groups that changes have been made and that management will avoid similar situations in future. A primary interest here may be in repairing damage to reputation, as in BP's case. Exxon's experience with the Alaskan oil spill from the Exxon Valdez shows how the task of restoring reputation continues long after the crisis.

Crisis Management and Strategies for Dealing with Crisis

Techniques for Anticipating Crises, and Preparations for Dealing With Them

Senior managers can call on the advice of specialist groups to help them to anticipate crisis situations. These will include management consultants skilled in futures research, scenario planning, and risk assessment; internal and external groups that take a strategic perspective on the organization's interests; and specialists such as public relations and public affairs practitioners. These practitioners have expertise in dealing with issues and crisis management, communication during a crisis, and reputation management.

The techniques such advisors will draw on include the following.

- *Futures research*, which makes use of techniques such as scenario planning—a planning approach that looks into the future in a disciplined way, creating and examining plausible scenarios—or Delphi studies. The latter work through a series of rounds of questions to experts on likely developments and possible risks. Questioning is pursued with, and summarized to, the group of experts until consensus, or consensus on differences of opinion, is reached. The results of Delphi studies can be used to inform decision-making.
- *Risk assessments*, which weigh the likelihood of events or situations developing and their possible impacts, and making judgments on threats and vulnerability. A specialization in risk assessment relates to reputation risk assessment—what could happen to seriously damage the organization's reputation, and what steps should be taken to avoid damage to reputation.
- *Issues management*, where topics of concern to particular groups are identified and tracked through environmental scanning and monitoring. Involved here may be social research, analysis of traditional and social media content, and close attention to the concerns of special interest groups.

Benefits for Routine Management of Attention to Crisis Management

Advisors on crisis management, whether internal or external to the organization, will encounter scepticism from management regarding the time, effort, and resources to be committed to crisis management. Too often, organizations have to experience a crisis situation before they will take the task of crisis management seriously—and for some this will be too late.

A benefit of attention to crisis management that may be overlooked is that it improves routine management. By asking at the planning stage what could go wrong, current practices and processes can be reexamined from this perspective. Improvements can be made that will avoid problems, as well as improve management perspectives, skills, and decision-making.

The Importance of Psychological Preparation

Attention to crisis management, to crisis planning, and to rehearsal of the response to potential crisis situations also improves management capability. Management is essentially an attempt to bring order to the work of groups of people and to the use of resources to achieve objectives. Crisis situations disrupt the existing order—sometimes catastrophically (think for example of the 9/11 terrorist attacks in New York in 2001, or the Japanese earthquake and tsunami in 2011)—and make exceptional demands on decision-makers. Managers and decision-makers, prepared through training for these situations, will still find them very difficult to deal with, but they will feel psychologically prepared for them. After BP's experience in the Gulf of Mexico, the company's CEO Tony Hayward told a BBC documentary program that he had felt ill-prepared for the demands of the situation.

Training for crisis management involves the following.

- Working through simulations. Simulations set up situations as realistically as possible, through the use of real-time but simulated developments, actors, and creation of pressure on decision-makers. They allow decision-makers, working through developed plans, to see how the plans—and they themselves—perform against the demands of the simulated situation.
- Preparation for the communication requirements of crisis situations. Who will speak for the organization, what will they be able to say, and how will they cope with media interest in crisis situations? The training here will be very specific—how can individual managers be prepared for the encounter with the media? How should they present themselves, what should they be prepared to say and where, and how should they be prepared for the approaches to questioning that will be taken by the media?

Summary and Further Steps

- The reality of the modern world is that there are endless possibilities for things to go badly wrong for organizations, large and small. They confront scarcity of resources, political uncertainty, heavy social expectations, competition, and potential for conflict.
- It is essential that management groups prepare for difficult situations. These will be defined organization by organization, depending on their interests, their relationships, and their prospects.
- Preparation must be thorough and undertaken by the most senior and responsible managers in the organization, since they have the authority to see that the preparations made will be followed through.
- Crisis management depends on management accepting the necessity of preparation for potential crisis. This means lifting attention from the immediate, the short term, and day-to-day practical concerns to look to the future, and to anticipate disruption.
- The phase of crisis planning lays the basis for all the work that will be done in managing a crisis, and this depends on a commitment of effort by senior management.
- The other essential of crisis management is the psychological preparation of managers for the demands it will make of them. This adds to management capability and is a grounding for management in the modern world.

Crisis Management and Strategies for Dealing with Crisis

More Info

Books:

Griffin, Andrew. *New Strategies for Reputation Management: Gaining Control of Issues, Crises & Corporate Social Responsibility*. London: Kogan Page, 2007.

Regester, Michael and Judy Larkin. *Risk Issues and Crisis Management in Public Relations: A Casebook of Best Practice*. 4th ed. London: Kogan Page, 2008.

Janis, Irving. *Group Think*. 2nd ed. Boston, MA: Houghton Mifflin, 1982.

Larkin, Judy. *Strategic Reputation Risk Management*. Basingstoke, UK: Palgrave Macmillan, 2002.

Lerbinger, Otto. *The Crisis Manager: Facing Risk and Responsibility*. Oxford, UK: Routledge Communication Series, 1997.

Movie:

Petersen, Wolfgang (dir). *The Perfect Storm*. Warner Bros, 2000.

Fraud: Minimizing the Impact on Corporate Image
by Tim Johnson
Regester Larkin, UK

This Chapter Covers

- Fraud is a threat faced by all organizations, regardless of their size or sector, that can easily plunge any organization into crisis, real or perceived.
- The key to crisis management—particularly when trust in business remains very low—is to set the agenda, communicate robustly, and not allow speculation or rumor to run rife.
- Robust communication strategies require organizations to consider their *message*, their *audience*, and the *medium* they will use to communicate their message.
- In cases of fraud, such messages should center on *concern, control, commitment,* and *containment*.

Introduction

The threat of fraud is faced by all organizations regardless of their size or sector. From the perspective of reputation management, controlling the impact of fraud is particularly challenging for two reasons:

1. That an organization has become a victim of fraud suggests either that someone in the organization is corrupt, or that the organization and its compliance systems are vulnerable. Neither possibility inspires confidence or trust.
2. The word "fraud" has a wide range of meanings. It can refer to a sustained, systemic failure that can bring an organization to its knees. Or it can refer to low-level compliance failure that, while regrettable, is unlikely to lead to long-lasting damage.

If fraud has been committed or is suspected, how can an organization's reputation be protected? First, we need to understand what reputation is and its importance. We also need to understand the rudiments of crisis reputation management.

Reputation and Why It Is Important

Reputation is hard to define and there are, consequently, many different definitions. Put simply, it is the sum total of what our stakeholders feel about a company and how they act as a result of that feeling. This sounds woolly, and indeed it is. Over the years, many attempts have been made to try and measure organizational reputation in quantifiable and, preferably, hard financial terms. Some progress has been made. But you still won't find a line on the asset—or liability—side of your balance sheet that refers to your organization's reputation.

Risk Management in an Uncertain World

Most practitioners and academics now accept that reputation will always be difficult to define and quantify. However, there is broad agreement that reputation is built on the trust stakeholders have in an organization, and that trust is far from woolly. On the contrary, trust brings hard commercial benefits: it helps to build strong brands, launch new products, secure licensing deals, recruit the best staff, and avoid intrusive regulation. Few would disagree that protecting that trust, and thus reputation, is critical to the business.

However, that's easier said than done because trust is a rare commodity—particularly in light of high-profile incidents, such as the rogue trading which led to the collapse of Barings Bank and, more recently, huge trading losses that have cost JP Morgan, Morgan Stanley, Société Générale, and UBS. In 2011, Ipsos MORI found that only 29% of those surveyed in the United Kingdom trusted business leaders to tell the truth. This lack of trust manifests itself in many ways, including a surge in the numbers of nongovernmental organizations, a breakdown in accepted societal structures, and the growth of antiglobalization sentiment that is often fueled by an aggressive 24/7 media. Even during times of "business as usual," reputation management is not an easy business.

So what should be done to protect organizational reputation during a crisis prompted by, for example, a case of fraud?

Crisis Communications

When something goes wrong, the natural instinct is to want to fix the problem behind closed doors. This is perfectly understandable, and in an ideal world the issue would be attended to and the relevant stakeholders told about the actions taken to rectify the problem—if anyone needs to be told at all.

However, in a world of citizen journalists, WikiLeaks, and social networking, even problems such as *suspected* fraud become harder to contain within an organization. News often leaks to the wider world long before the organization has found a solution. Sometimes, news can reach the outside world even before it reaches management.

In such circumstances, the key to reputation management is to be ready and willing to communicate about the problem, outlining what has happened, the extent of the situation, and, critically, what the organization is doing to put it right. The organization must establish itself as the authoritative source of information about the situation, crushing harmful speculation and robustly deflecting the vicious rumors that inevitably accompany such stories.

In developing such a communication plan, an organization needs to consider the following factors:

- *Message:* What it will say about the situation and when.
- *Audience:* Who it will say it to and in what order.
- *Medium:* The platform it will use to say it.

Each crisis situation is different, but in cases of fraud organizations should consider the following.

Fraud: Minimizing the Impact on Corporate Image

Key Considerations
Messaging
Fraud can be brought to an organization's attention in many ways (for example, internal audit, whistleblower, media inquiry, etc.). Regardless of how the news reaches an organization, holding messages are required immediately. These are for use until an investigation is complete.

In cases of fraud, the "4Cs" should be applied:

- *Concern* (for what's happened): The incident is being treated extremely seriously.
- *Control* (of the situation): The claim is being investigated thoroughly.
- *Containment* (of the consequences): While regrettable, this will not have a material impact on the organization.
- *Commitment* (to compliance): If this is the first time such an allegation has been made, initiate organizational compliance procedures over and above what is required.

It is often helpful when considering how to communicate *containment* to try to contextualize the message. For example, "This allegation relates to less than 0.0001% of turnover in just one of 30 markets we operate in." However, it is important not to downplay the alleged fraud if it may materially affect the organization. Also, until an investigation is complete, it can be dangerous to second guess the findings.

Audience
Although an organization should be prepared to use this interim messaging widely and rapidly, if news of the alleged fraud is successfully contained, it may only be necessary to communicate it to a limited number of stakeholders.

To help identify which stakeholders should be proactively notified with these messages, an organization must consider which of its audiences have an interest in the problem (often the same as those who are affected) and those who may have some influence on its resolution. This can be plotted on a simple matrix (see Figure 1).

Figure 1. Stakeholder mapping—Matrix to plot which audiences should receive proactive communication in a crisis

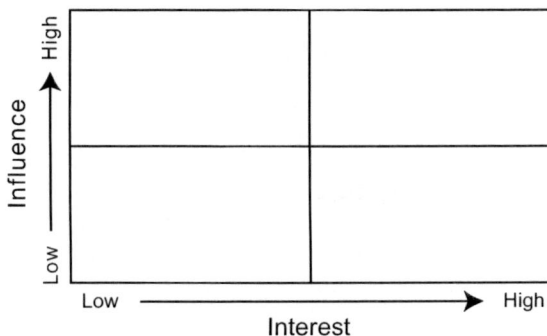

Risk Management in an Uncertain World

Stakeholders in the top right-hand corner should be identified and a stepwise briefing process should be implemented. Clearly, this takes judgment and this is where external reputation management advisers can prove extremely valuable. When identifying your stakeholders, it is important to be as precise about them as possible:

- "Staff" is not a useful stakeholder category. A specific level of management in a specific department is more useful and will focus the process.
- There may be regulatory procedures to follow (for example, stock market announcements), and these must be adhered to.

Finally, it is worth noting that through an organization's ongoing "peacetime" reputation management program, solid relationships with these key stakeholders should already be in place. This is known as banking "relationship credit." The more relationship credit you have banked in peacetime, the easier it will be to draw on that credit in times of crisis, and the more forgiving stakeholders are likely to be.

Medium

Organizations should form a small senior team to manage the situation and use it to brainstorm how these messages will be delivered—for example, in person, by written letter, or via the media. However, it is important not to overcomplicate this part of the process. It is simply a case of considering those who are being notified and thinking about how they might be approached. For example, the head of a regulatory agency may appreciate it if the organization's CEO/MD delivers these messages in person.

There are three additional considerations:

- It is often useful (with legal advice) to put things in writing to stakeholders. However, always assume that whatever is written may be leaked or will be subject to compulsory disclosure in court. As a rule, an organization should not write anything that it wouldn't want to see in a newspaper or hear repeated in front of a judge.
- Putting a human face on things should never be underestimated. Even hardened regulators and authorities respond better to a one-to-one interaction than they do to a statutory written communication. An organization should consider who should make that interaction. If a minor regulatory infringement is involved, using the CEO to deliver the message is not appropriate. But if potentially it's a major issue, the chair of the board is the only appropriate person.
- Everyone will have their own agenda. The regulators and authorities may decide that they want to showcase an organization or defend any possible allegations that they were "asleep at the wheel." It can't stop the latter, but it is better that those stakeholders know your side of the story.

Drawing a Line Under the Situation

Once the investigation has been completed, you should be prepared to draw a line under the situation. To some extent, doing this depends on how public the situation has become. If developments have been highly publicized and commented on by a wide range of stakeholders, then a wide-ranging outreach plan should be developed

Fraud: Minimizing the Impact on Corporate Image

to communicate how the organization intends to move on. If the situation has been relatively contained, the outreach plan may be far less reaching.

Irrespective of the reach of the communications plan, some underpinning messages will be required. The 4Cs formula outlined above can be revisited and revised. And it is important that messages are backed up by action. With stakeholders now less trusting than they were, an organization needs to *show* its audience that it has moved on, not just *tell* them.

For example, if an organization says that it is committed to compliance, can it allow the person accused of fraud to stay in their position? If an organization says it is confident that the situation is now under control and cannot recur, what tangible steps or changes can it point to as evidence that it really has acted to prevent a recurrence?

The more tangible the evidence underpinning the message, the firmer and more convincing the line the organization will be able to draw under the situation.

Case Studies

Société Générale

In January 2008, French bank Société Générale, one of Europe's biggest financial services companies, revealed that it had lost €4.9 billion in an incident of fraud involving a single futures trader.

Société Générale managed the incident very well. It was required to respond quickly, and it did: Two days after suspicions were aroused concerning unusual trading activity, the bank's chairman, Daniel Bouton, informed the governor of the Bank of France and suspended the trader in question, Jérôme Kerviel. The company successfully contained the incident and did not attempt to play down its potential magnitude.

As soon as Société Générale had complied with regulatory reporting, it moved from interim messages to release its first public statement, establishing itself as the authoritative source of information. The bank successfully communicated its containment of the crisis—despite admitting that it would need significant new capital to offset the losses, it reassured the financial community that it was still on course to make a good profit. It continued to give information to the authorities before releasing a candid statement about the incident: who Kerviel was, arbitrage activities, the method behind the fraud, how it was uncovered, and measures taken since the event.

The crisis required a human face, and the frontline response came from Daniel Bouton, whose resignation as executive chairman was rejected by the board early on but who eventually stepped down to nonexecutive chairman.

At the time of writing, Jérôme Kerviel has a pending appeal of his conviction from January 2008 for breach of trust, computer abuse, and falsification. He has denied any wrongdoing and claims that the bank knew of his actions but let him continue as long as he was making money. He has filed two lawsuits against Société Générale which the bank has countersued. The on-going investigation, political scrutiny by the new Socialist French government, and a €4 million fine imposed by France's banking regulator, clearly makes it difficult for Société Générale to draw a line under the incident. However, the bank has laid the groundwork with strong actions to back up its messages (such as internal investigations into compliance) that have demonstrated its determination to stakeholders and built trust in the process.

Cobalt Energy
In March 2011 the US Securities and Exchange Commission (SEC) began an informal investigation into the US exploration company Cobalt Energy, following allegations of a connection between one of Cobalt's local partners in Angola and Angolan government officials. US companies are subject to US laws, including the Foreign Corrupt Practices Act (FCPA), regardless of where they operate.

Cobalt took a proactive approach and began its own internal probe into the issue. By the time the SEC had issued a formal order of investigation in November 2011, the company was prepared to respond and contacted the US Department of Justice to offer to respond to any requests. In February 2012 Cobalt Energy alerted its shareholders to the risk of liabilities under anti-corruption laws in the United States and in a regulatory filing stated its activities had been in compliance with anti-bribery laws.

The pressure on Cobalt Energy was amplified when a report published by the *Financial Times* newspaper in April 2012 alleged that three very senior Angolan officials had held concealed interests in the company's venture. This drew further public attention to the issue and as much as US$1.4 billion (almost 11%) was wiped off Cobalt Energy's market capitalization in one day.

Cobalt responded quickly to the media coverage, refuting allegations of wrongdoing and disputing the Financial Times' sources. The company continued to monitor coverage of the issue as demonstrated by a formal letter sent to the *Foreign Policy Journal* in response to an article in May 2012.

At the time of writing, Cobalt Energy remains under investigation yet committed to its operations in Angola. Its share price, however, has fluctuated dramatically since it reached an all-time high earlier in the year, and has been down by 30%.

Cobalt Energy's case illustrates the importance of due diligence in new market entry and how operating in countries with a reputation for corrupt practices can bring public allegations of suspected fraud, damage a company's corporate reputation, significantly impact its financial valuation, and threaten its license to operate.

Conclusion
Everyone accepts that things go wrong from time to time. What most organizations will be judged on is not that something has gone wrong, but on how they respond to the situation.

Although every situation has its own dynamics, by following some of the broad guidelines outlined above organizations that are victims of fraud will go a long way toward protecting themselves from some of the reputational fallout they may suffer. Ultimately these guidelines should also help to maintain that all-important trust from their stakeholders.

> **More Info**
>
> **Books:**
> Alsop, Ronald J. *The 18 Immutable Laws of Corporate Reputation: Creating, Protecting and Repairing Your Most Valuable Asset*. London: Kogan Page, 2006.
> Doorley, John, and Helio F. Garcia. *Reputation Management: The Key to Successful Corporate and Organizational Communication*. New York: Routledge, 2005.
> Griffin, Andrew. *New Strategies for Reputation Management: Gaining Control of Issues, Crises and Corporate Social Responsibility*. London: Kogan Page, 2007.
> Larkin, Judy. *Strategic Reputation Risk Management*. Basingstoke, UK: Palgrave Macmillan, 2003.
> Mitroff, Ian. *Why Some Companies Emerge Stronger and Better from a Crisis*. New York: Amacom, 2005.
> O'Hanlon, Bill. *Thriving Through Crisis: Turn Tragedy and Trauma Into Growth and Change*. New York: Perigee, 2005.
> Regester, Michael, and Judy Larkin. *Risk Issues and Crisis Management in Public Relations: A Casebook of Best Practice*. 4th ed. London: Kogan Page, 2008.
> Ulmer, Robert, Timothy L. Sellnow, and Matthew W. Seeger. *Effective Crisis Communication: Moving from Crisis to Opportunity*. Thousand Oaks, CA: Sage Publications, 2006.
> van Reil, Cees B. M., and Charles J. Fombrun. *Essentials of Corporate Communication: Implementing Practices for Effective Reputation Management*. New York: Routledge, 2006.
>
> **Articles:**
> Ettenson, Richard, and Jonathan Knowles. "Don't confuse reputation with brand." *MIT Sloan Management Review* 49:2 (2008). Online at: tinyurl.com/c8y6mhc
> Gardberg, N., and C. Fombrun. "The global reputation quotient project: First steps towards a cross-nationally valid measure of corporate reputation." *Corporate Reputation Review* 4:4 (2002): 303–307. Online at: dx.doi.org/10.1057/palgrave.crr.1540151
> MacMillan, Keith, Kevin Money, Steve Downing, and Carola Hillenbrand. "Giving your organisation SPIRIT: An overview and call to action for directors on issues of corporate governance, corporate reputation and corporate responsibility." *Journal of General Management* 30:2 (Winter 2005): 15–42.

Risk in the External Environment

Measuring Company Exposure to Country Risk
by Aswath Damodaran
Stern School of Business, New York University, USA

This Chapter Covers

» We focus on a question: Once we have estimated a country risk premium, how do we evaluate a company's exposure to country risk?
» In the process, we will argue that a company's exposure to country risk should not be determined by where it is incorporated and traded.
» By that measure, neither Coca-Cola nor Nestlé are exposed to country risk. Exposure to country risk should come from a company's operations, making country risk a critical component of the valuation of almost every large multinational corporation.

Introduction

If we accept the proposition of country risk, the next question that we have to address relates to the exposure of individual companies to country risk. Should all companies in a country with substantial country risk be equally exposed to country risk? While intuition suggests that they should not, we will begin by looking at standard approaches that assume that they are. We will follow up by scaling country risk exposure to established risk parameters such as betas (β), and complete the discussion with an argument that individual companies should be evaluated for exposure to country risk.

The Bludgeon Approach

The simplest assumption to make when dealing with country risk, and the one that is most often made, is that all companies in a market are equally exposed to country risk. The cost of equity for a firm in a market with country risk can then be written as:

Cost of equity = Risk-free rate + β (Mature market premium) + Country risk premium

Thus, for Brazil, where we have estimated a country risk premium of 4.43% from the melded approach, each company in the market will have an additional country risk premium of 4.43% added to its expected returns. For instance, the costs of equity for Embraer, an aerospace company listed in Brazil, with a beta[1] of 1.07 and Embratel, a Brazilian telecommunications company, with a beta of 0.80, in US dollar terms would be:

Cost of equity for Embraer = 3.80% + 1.07 (4.79%) + 4.43% = 13.35%

Cost of equity for Embratel = 3.80% + 0.80 (4.79%) + 4.43% = 12.06%

Note that the risk-free rate that we use is the US treasury bond rate (3.80%), and that the 4.79% figure is the equity risk premium for a mature equity market (estimated from historical data in the US market). It is also worth noting that analysts estimating the cost of equity for Brazilian companies, in US dollar terms, often use the Brazilian ten-year dollar-denominated rate as the risk-free rate. This is dangerous, since it is often also accompanied with a higher risk premium, and ends up double counting risk.

Risk Management in an Uncertain World

The Beta Approach

For those investors who are uncomfortable with the notion that all companies in a market are equally exposed to country risk, a fairly simple alternative is to assume that a company's exposure to country risk is proportional to its exposure to all other market risk, which is measured by the beta. Thus, the cost of equity for a firm in an emerging market can be written as follows:

$$\text{Cost of equity} = \text{Risk-free rate} + \beta\,(\text{Mature market premium} + \text{Country risk premium})$$

In practical terms, scaling the country risk premium to the beta of a stock implies that stocks with betas above 1.00 will be more exposed to country risk than stocks with a beta below 1.00. For Embraer, with a beta of 1.07, this would lead to a dollar cost of equity estimate of:

$$\text{Cost of equity for Embraer} = 3.80\% + 1.07\,(4.79\% + 4.43\%) = 13.67\%$$

For Embratel, with its lower beta of 0.80, the cost of equity is:

$$\text{Cost of equity for Embratel} = 3.80\% + 0.80\,(4.79\% + 4.43\%) = 11.18\%$$

The advantage of using betas is that they are easily available for most firms. The disadvantage is that while betas measure overall exposure to macroeconomic risk, they may not be good measures of country risk.

The Lambda Approach

The most general, and our preferred, approach is to allow for each company to have an exposure to country risk that is different from its exposure to all other market risk. For lack of a better term, let us term the measure of a company's exposure to country risk to be lambda (λ). Like a beta, a lambda will be scaled around 1.00, with a lambda of 1.00 indicating a company with average exposure to country risk and a lambda above or below 1.00 indicating above or below average exposure to country risk. The cost of equity for a firm in an emerging market can then be written as:

$$\text{Expected return} = R_f + \beta\,(\text{Mature market equity risk premium}) + \lambda\,(\text{Country risk premium})$$

Note that this approach essentially converts our expected return model to a two-factor model, with the second factor being country risk, with λ measuring exposure to country risk.

Determinants of Lambda

Most investors would accept the general proposition that different companies in a market should have different exposures to country risk. But what are the determinants of this exposure? We would expect at least three factors (and perhaps more) to play a role.

1. *Revenue source:* The first and most obvious determinant is how much of the revenues a firm derives from the country in question. A company that derives 30% of its revenues from Brazil should be less exposed to Brazilian country risk than a company that derives 70% of its revenues from Brazil. Note, though, that this then opens up the possibility that a company can be exposed to the risk in many countries. Thus, the company that derives only 30% of its revenues from Brazil may derive its remaining revenues from

Measuring Company Exposure to Country Risk

Argentina and Venezuela, exposing it to country risk in those countries. Extending this argument to multinationals, we would argue that companies like Coca-Cola and Nestlé can have substantial exposure to country risk because so much of their revenues comes from emerging markets.

2. *Production facilities:* A company can be exposed to country risk, even if it derives no revenues from that country, if its production facilities are in that country. After all, political and economic turmoil in the country can throw off production schedules and affect the company's profits. Companies that can move their production facilities elsewhere can spread their risk across several countries, but the problem is exaggerated for those companies that cannot move their production facilities. Consider the case of mining companies. An African gold mining company may export all of its production but it will face substantial country risk exposure because its mines are not movable.

3. *Risk management products:* Companies that would otherwise be exposed to substantial country risk may be able to reduce this exposure by buying insurance against specific (unpleasant) contingencies and by using derivatives. A company that uses risk management products should have a lower exposure to country risk—a lower lambda—than an otherwise similar company that does not use these products.

Ideally, we would like companies to be forthcoming about all three of these factors in their financial statements.

Measuring Lambda

The simplest measure of lambda is based entirely on revenues. In the last section, we argued that a company that derives a smaller proportion of its revenues from a market should be less exposed to country risk. Given the constraint that the average lambda across all stocks has to be 1.0 (someone has to bear the country risk!), we cannot use the percentage of revenues that a company gets from a market as lambda. We can, however, scale this measure by dividing it by the percentage of revenues that the average company in the market gets from the country to derive a lambda.

$$(\lambda_j = \% \text{ of revenue in country}_{Company}) \div \% \text{ of revenue in country}_{Average\ company\ in\ market}$$

Consider the two large and widely followed Brazilian companies—Embraer, an aerospace company that manufactures and sells aircraft to many of the world's leading airlines, and Embratel, the Brazilian telecommunications giant. In 2002, Embraer generated only 3% of its revenues in Brazil, whereas the average company in the market obtained 85% of its revenues in Brazil.[2] Using the measure suggested above, the lambda for Embraer would be:

$$\lambda_{Embraer} = 3\% \div 85\% = 0.04$$

In contrast, Embratel generated 95% of its revenues from Brazil, giving it a lambda of

$$\lambda_{Embratel} = 95\% \div 85\% = 1.12$$

Following up, Embratel is far more exposed to country risk than Embraer and will have a much higher cost of equity.

Risk Management in an Uncertain World

The second measure draws on the stock prices of a company and how they move in relation to movements in country risk. Bonds issued by countries offer a simple and updated measure of country risk; as investor assessments of country risk become more optimistic, bonds issued by that country go up in price, just as they go down when investors become more pessimistic. A regression of the returns on a stock against the returns on a country bond should therefore yield a measure of lambda in the slope coefficient. Applying this approach to Embraer and Embratel, we regressed monthly stock returns on the two stocks against monthly returns on the ten-year dollar-denominated Brazilian government bond and arrived at the following results:

$$\text{Return}_{\text{Embraer}} = 0.0195 + 0.2681\, \text{Return}_{\text{Brazil dollar-bond}}$$

$$\text{Return}_{\text{Embratel}} = -0.0308 + 2.0030\, \text{Return}_{\text{Brazil dollar-bond}}$$

Based upon these regressions, Embraer has a lambda of 0.27 and Embratel has a lambda of 2.00. The resulting dollar costs of equity for the two firms, using a mature market equity risk premium of 4.79% and a country equity risk premium of 4.43% for Brazil are:

Cost of equity for Embraer = 3.80% + 1.07 (4.79%) + 0.27 (4.43%) = 10.12%

Cost of equity for Embratel = 3.80% + 0.80 (4.79%) + 2.00 (4.43%) = 16.49%

What are the limitations of this approach? The lambdas estimated from these regressions are likely to have large standard errors; the standard error in the lambda estimate of Embratel is 0.35. It also requires that the country have bonds that are liquid and widely traded, preferably in a more stable currency (dollar or euro).

Risk Exposure in Many Countries
The discussion of lambdas in the last section should highlight a fact that is often lost in valuation. The exposure to country risk, whether it is measured in revenues, earnings, or stock prices, does not come from where a company is incorporated but from its operations. There are US companies that are more exposed to Brazilian country risk than is Embraer. In fact, companies like Nestlé, Coca-Cola, and Gillette have built much of their success on expansion into emerging markets. While this expansion has provided them with growth opportunities, it has also left them exposed to country risk in multiple countries.

In practical terms, what does this imply? When estimating the costs of equity and capital for these companies and others like them, we will need to incorporate an extra premium for country risk. Thus, the net effect on value from their growth strategies will depend upon whether the growth effect (from expanding into emerging markets) exceeds the risk effect. We can adapt the measures suggested above to estimate the risk exposure to different countries for an individual company.

We can break down a company's revenue by country and use the percentage of revenues that the company gets from each emerging market as a basis for estimating lambda in that market. While the percentage of revenues itself can be used as a lambda, a more precise estimate would scale this to the percentage of revenues that the average company in that market gets in the country.

Measuring Company Exposure to Country Risk

If companies break earnings down by country, these numbers can be used to estimate lambdas. The peril with this approach is that the reported earnings often reflect accounting allocation decisions and differences in tax rates across countries.

If a company is exposed to only a few emerging markets on a large scale, we can regress the company's stock price against the country bond returns from those markets to get country-specific lambdas.

Conclusion

A key issue, when estimating costs of equity and capital for emerging market companies relates to how this country risk premium should be reflected in the costs of equities of individual companies in that country. While the standard approaches add the country risk premium as a constant to the cost of equity of every company in that market, we argue for a more nuanced approach where a company's exposure to country risk is measured with a lambda. This lambda can be estimated either by looking at how much of a company's revenues or earnings come from the country—the greater the percentage, the greater the lambda—or by regressing a company's stock returns against country bond returns—the greater the sensitivity, the higher the lambda. If we accept this view of the world, the costs of equity for multinationals that have significant operations in emerging markets will have to be adjusted to reflect their exposure to risk in these markets.

More Info

Book:
Falaschetti, Dominic, and Michael Annin Ibbotson (eds). *Stocks, Bonds, Bills and Inflation*. Chicago, IL: Ibbotson Associates, 1999.

Articles:
Booth, Laurence. "Estimating the equity risk premium and equity costs: New ways of looking at old data." *Journal of Applied Corporate Finance* 12:1 (Spring 1999): 100–112. Online at: dx.doi.org/10.1111/j.1745-6622.1999.tb00665.x

Chan, K. C., G. Andrew Karolyi, and René M. Stulz. "Global financial markets and the risk premium on US equity." *Journal of Financial Economics* 32:2 (October 1992): 137–167. Online at: dx.doi.org/10.1016/0304-405X(92)90016-Q

Damodaran, Aswath. "Country risk and company exposure: Theory and practice." *Journal of Applied Finance* 13:2 (Fall/Winter 2003): 64–78.

Godfrey, Stephen, and Ramon Espinosa. "A practical approach to calculating the cost of equity for investments in emerging markets." *Journal of Applied Corporate Finance* 9:3 (Fall 1996): 80–90. Online at: dx.doi.org/10.1111/j.1745-6622.1996.tb00300.x

Indro, Daniel C., and Wayne Y. Lee. "Biases in arithmetic and geometric averages as estimates of long-run expected returns and risk premium." *Financial Management* 26:4 (Winter 1997): 81–90. Online at: www.jstor.org/stable/3666130

Stulz, René M. "Globalization, corporate finance, and the cost of capital." *Journal of Applied Corporate Finance* 12:3 (Fall 1999): 8–25.
Online at: dx.doi.org/10.1111/j.1745-6622.1999.tb00027.x

Report:
Damodaran, Aswath. "Measuring company risk exposure to country risk: Theory and practice." September 2003. Online at: tinyurl.com/77ozxll [PDF].

Notes

1. We used a bottom-up beta for Embraer, based upon an unlevered beta of 0.95 (estimated using aerospace companies listed globally) and Embraer's debt-to-equity ratio of 19.01%.
2. To use this approach, we need to estimate the percentage of revenues both for the firm in question and for the average firm in the market. While the former may be simple to obtain, estimating the latter can be a time-consuming exercise. One simple solution is to use data that are publicly available on how much of a country's gross domestic product comes from exports. According to the World Bank data in this table, Brazil got 23.2% of its GDP from exports in 2008. If we assume that this is an approximation of export revenues for the average firm, the average firm can be assumed to generate 76.8% of its revenues domestically. Using this value would yield slightly higher betas for both Embraer and Embratel.

The Impact of Climate Change on Business
by Graham Dawson
University of Buckingham, UK

This Chapter Covers

- The impact of climate change on business—or the monetary value of the costs that may be incurred by affected parties and the benefits that they may accrue— is difficult to assess with any degree of precision.
- The Stern Review and the United Nations Intergovernmental Panel on Climate Change (IPCC) have reported the results of running complex computer models that integrate climate science and economics with the aim of predicting the economic impact of climate change far into the future.
- There is no agreement concerning the appropriate discount rate or the monetary value of effects where market prices are not available.
- Uncertainty also surrounds the rate, and carbon-intensiveness, of the growth of the world economy for decades and even centuries ahead, while the hypothesis of anthropogenic climate change itself continues to be controversial.

The Global Impact of Climate Change on People

The standard approach to assessing the economic impact of climate change on business requires giving a monetary value to the costs that may be incurred by those affected and the benefits that may accrue to them.

The most comprehensive attempt to do this is the Stern Review (2007), commissioned by the UK government, which predicts severe impacts from an average global temperature rise of 2–3°C within the next 50 years or so. These impacts include an increased risk of flooding from melting glaciers, followed by disruption to water supplies, affecting up to one-sixth of the world's population, mainly in the Indian subcontinent and parts of China and South America. In higher-latitude areas, such as Northern Europe, agricultural yields may increase with a temperature increase of 2–3°C, but declining yields, especially in Africa, could leave hundreds of millions of people without sufficient food. Increased mortality from heat-related deaths and the spread of tropical diseases is predicted, although there will be fewer deaths from exposure to cold. With warming of 3–4°C, thermal expansion of the oceans is predicted to cause rising sea levels, which could lead to inundation of low-lying coastal land, displacing "tens to hundreds of millions" of people. The risks are greatest for Southeast Asia (Bangladesh and Vietnam), small islands in the Caribbean and the Pacific, and large coastal cities, such as Tokyo, New York, Cairo, and London. Extreme weather events may become more frequent.

Risk Management in an Uncertain World

> **Case Study**
>
> ### What Would This Mean for Business Activity in, for example, the United States?
>
> If predictions such as those reported by Stern prove to be accurate, business will be forced to adapt to changes in climate. Adaptation would involve a range of measures of varying cost. In the United States, temperature increases of up to 2–3°C might cause the wheat belt to shift northward into Canada; US farmers in the Midwest would have to plant new crop varieties, a fairly routine adjustment. In northern areas, winter deaths from exposure to the cold would fall and tourism might increase. Further south, the melting of snow could make the water supply to California and the Mississippi basin more erratic, causing more acute problems for agriculture. Deaths from exposure to heat and the cost of air conditioning and refrigeration would increase. At higher temperatures, southern parts of the United States would see an increased risk of extreme weather events, requiring substantial investment to defend low-lying cities, such as New Orleans and New York, from flooding.

Modeling the Costs of Climate Change

It is easy enough to put a monetary value on some of these impacts. For example, there is a lot of expensive real estate with known market prices in major coastal cities such as London, New York, and Tokyo. Moreover, without offices or factories for people to work in, or homes for them to live in, output would fall, at least for a while. Declining crop yields (adjusted for higher prices) and also fish stocks would reduce the value of world output. Standard practice is to estimate the loss of output consequent upon people's incapacity for paid and unpaid work.

Quantifying these predicted impacts of climate change in monetary terms requires degrees of certainty and precision that may not be attainable. Both the science and the economics of climate change are subject to considerable uncertainty and are therefore deeply controversial.

The impacts of climate change on business depend on the magnitude of temperature changes associated with different concentrations of CO_2 and other greenhouse gas (GHG) emissions, according to the scientific hypothesis of anthropogenic climate change. The earliest studies of the economic impact of climate change assumed a doubling of atmospheric concentrations of CO_2 by 2050, and estimated the costs of the resulting increase in global mean surface temperature at approximately 2% of world gross domestic product (GDP).

Subsequent modeling of the economic impact of climate change has sought to integrate scientific models of the global climate and economic models of future world economic growth. The anthropogenic hypothesis holds that most of the observed rise in temperature has been caused by GHG emissions from fossil fuel use in economic activity. The future path of GHG emissions depends on the rate of growth of world economic activity and how that growth is divided between more and less carbon-intensive processes. So predicting the future path of GHG emissions, and hence the impact of climate change on business, involves modeling the rate of growth of the world economy well into the future.

The Impact of Climate Change on Business

The United Nations Intergovernmental Panel on Climate Change (IPCC) occupies a near-monopoly position in disseminating climate science to policy makers throughout the world. It does not predict future temperature increases and their impacts but prepares a number of illustrative outcomes, using integrated assessment models (IAM). Models of world economic growth and consequent GHG emissions are combined with climate science models, showing the links between those GHG emissions and temperature change.

The Stern Review used PAGE2002, an IAM designed by the UK government in 2000 and modified two years later. Stern claims that the overall costs and risks of business-as-usual (BAU) climate change would be equivalent to losing 5–20% of global GDP each year, "now and forever," but this may not be as apocalyptic as it sounds.

Stern explains the different stages by which this estimate of the economic impact of climate change was reached. The model is run to simulate a period of 200 years or more and "produces a mean warming of 3.9°C relative to pre-industrial in 2100." The first stage indicates that the costs and risks of climate change that can be quantified in terms of market values (basically, lost output) would be equivalent to losing at least 5% of global GDP each year, "now and forever."

At this point, Stern departs from most other models by adding in "nonmarket" impacts on the environment and human health. Nonmarket impacts are those that cannot be given a monetary value by referring to a market price (for instance, the price of land lost to coastal flooding). The costs of disease or of lost agricultural land in subsistence economies, for example, do not have a market price. Including this second stage increases the total cost from 5% to 11% of global GDP. These estimates are highly controversial. Since standard practice is to estimate health impacts in terms of lost output from incapacity to work, applying this and other techniques to estimate the cost of nonmarket impacts is subject to considerable uncertainty. It has also been argued that the degree to which both disease and casualties from natural disasters are related to income rather than environmental factors is not taken into account.

The third stage adds amplifying feedback effects, including the risk of catastrophic climate change, which increase the potential total cost from 11% to 14% of global GDP. Finally, Stern considers the view that a disproportionate burden of climate change would fall on poor regions. If this were given a stronger relative weight, the total cost of global warming could increase to "around 20%" of global GDP. Stern arrives at such a large adjustment for poor regions because he assumes that vulnerability to climate change is independent of development, but it seems more likely that such vulnerability depends on the capacity to adapt and hence on the level of development.

Uncertainties in the Economic Valuation of Impacts
"Now and Forever"
The phrase "now and forever" invites examination. The effects of climate change are expected to occur year by year over a very long period of time. The Stern Review calculates the present value of the costs of climate change by averaging the total costs over the number of years the model runs at a rate of discount. Nordhaus ran the Stern model to calculate the costs of climate change, including nonmarket and catastrophic

impacts that take Stern's estimate up to 14% of world output, for each year the model covers. According to Nordhaus, the model projects a mean loss of only 0.4% of world output in 2060, rising to 2.9% in 2100 and 13.8% in 2200. Losses averaging about 1% over the period 2000–2100 become about 14% "now and forever" because the losses in the distant future are extremely high (and a low discount rate is used). Nordhaus argues that, "using the [Stern] *Review's* methodology, more than half of the estimated damages 'now and forever' occur after the year 2800."

Discounting

For most people, $100 is worth more today than $100 next year because there is a degree of uncertainty about what might happen between now and next year; they would prefer to have $100 to spend right now to having it at some point in an uncertain future. In other words, the *present value* of that $100 payable to you in ten years is less than $100 paid to you now. Similarly, the expected future costs, no less than the benefits, of an event or occurrence should be discounted, i.e. reduced in value, in order to estimate their present value.

Since many economic impacts of climate change are not expected to occur until decades or even centuries into the future, their occurrence is inevitably subject to a degree of uncertainty. The impacts of catastrophic climate change may never happen, so economists discount, or reduce the value of, their costs. As you add up the costs of climate change year by year, you might want to adjust downward those expected in later years—that is, you might want to *discount* them to reflect the uncertainty of their occurrence. The higher the rate at which you discount such costs, the lower will be their present value.

The discount rate used may influence the results of a model more than any other parameter or value used in the model. There is no agreement about the appropriate rate of discount to use, and Stern argues that any discount rate greater than zero unfairly devalues the interests of future generations. He sets the "pure time preference rate" at zero, on the grounds that a future generation has the same claim on our ethical attention as the current one. Based on a zero pure time preference rate, the discount rates used in Stern's running of PAGE2002 are lower than those used in most other models and do much to explain why Stern's "baseline" cost of 5% of world GDP is higher than the results of other models (typically 1–2% of world GDP). Other ethical approaches are at least as convincing. For example, agent-relative ethics holds that agents naturally value people who are linked to them by kinship or proximity above strangers who are remote in space or time. This approach implies a higher discount rate, which would reduce the loss from "business as usual" in Stern's model substantially below 5% of world GDP.

The estimate is an annual average for an indefinite future; losses are low for the first 50 years or so, and using unusually low discount rates produces a high present value for the catastrophic losses predicted for 2200 and beyond. By that time, given rates of world economic growth sufficient to cause the projected carbon emissions and climate change, it is reasonable to assume that most people will be very much better off than the current generation, although not quite as much better off as they would have been in the absence of climate change.

The Impact of Climate Change on Business

Scenarios of Future World Economic Growth
How much better off would these future generations be, and which groups of people would gain most? What will the world economy look like 100 years from now? Wisely, the IPCC has demurred from making any such prediction, offering instead six illustrative scenarios of possible future courses that the world economy might take. In 2007 the IPCC reported the "best estimates and likely ranges for global average surface air warming for six…emissions marker scenarios." The best estimate for the low scenario is 1.8°C, and the best estimate for the high scenario is 4.0°C. The important point here is that scenarios are descriptions of possible outcomes to which no probability can be attached. Of the six scenarios, the IPCC asserts that that: "All should be considered equally sound." If it is impossible to assess the risk of any of the associated impacts, there is radical uncertainty.

It is widely believed that the impact of an increase in global temperature of less than 2°C will be mild, and that cereal yields will actually increase in temperate regions. With a global temperature increase of 4°C, the impacts are projected to be catastrophic, with up to 80 million people exposed to malaria, and up to 300 million more affected by coastal flooding each year, with rising risks of extreme weather events. But, on the IPCC's own admission, it is impossible to say whether the impact of climate change will be mild or catastrophic.

Uncertainty in Climate Science
Uncertainty also surrounds the science of climate change. In its most recent report, the IPCC claims that there is 90% certainty that most of the increase in global mean temperature since the middle of the twentieth century has been caused by the observed increase in greenhouse gas concentrations in the atmosphere. This is actually a rather cautious and vague claim, because it is consistent with a significant role for natural causes being the reason for the rise in global temperature. In the years since 1998 global temperature has not risen, and critics of the IPCC argue that the scientific evidence for dangerous change is far from overwhelming.

Conclusion
The impact of climate change on business, or the monetary value on the costs that may be incurred by affected parties and the benefits that they may accrue, is difficult to assess with any degree of precision.

The Stern Review and IPCC have reported the results of running complex computer models that integrate climate science and economics with the aim of predicting the economic impact of climate change into the remote future. However, there is no agreement concerning (i) the appropriate discount rate and (ii) the monetary value of effects where market values are unavailable. Uncertainty also surrounds the rate, and carbon-intensiveness, of the growth of the world economy for decades and even centuries ahead, while the hypothesis of anthropogenic climate change itself continues to be controversial.

Making It Happen
Business may be affected by policies to mitigate climate change as much as by climate change itself. In the negotiations for the Kyoto Protocol, which seeks to establish

a global framework for reductions in GHG emissions, the fossil fuel producers and users resisted aggressive reductions, while insurance companies and renewable energy producers were more favorably disposed toward them. It is not clear whether aggressive mitigation policies will survive the financial crisis of 2008, with many policy makers more concerned to reduce the effects of the expected global recession than the more distant threats posed by climate change.

> **More Info**
>
> **Books:**
> Lawson, Nigel. *An Appeal to Reason: A Cool Look at Global Warming*. London: Duckworth, 2008.
> Nordhaus, William. *The Challenge of Global Warming: Economic Models and Environmental Policy*. New Haven, CT: Yale University Press, 2007.
> Online at: nordhaus.econ.yale.edu/dice_mss_072407_all.pdf
> Singer, S. Fred, and Dennis T. Avery. *Unstoppable Global Warming: Every 1500 Years*. Lanham, MD: Rowman & Littlefield, 2006.
> Stern, Nicholas. *The Economics of Climate Change: The Stern Review*. Cambridge, UK: Cambridge University Press, 2007.
>
> **Articles:**
> Beckerman, Wilfred, and Cameron Hepburn. "Ethics of the discount rate in the Stern Review." *World Economics* 8:1 (2007): 187–210. Online at: tinyurl.com/7adyfc8
> Brittan, Samuel. "On climate change and good sense." *Financial Times* (February 9, 2007). Online at: www.samuelbrittan.co.uk/text268_p.html
> Byatt, Ian, Ian Castles, Indur M. Goklany, David Henderson, *et al*. "The Stern Review: A dual critique. Part II: Economic aspects." *World Economics* 7:4 (2006): 199–232. Online at: tinyurl.com/7dsdfrr
> Carter, Robert M., C. R. de Freitas, Indur M. Goklany, David Holland, *et al*. "The Stern Review: A dual critique. Part I: The science." *World Economics* 7:4 (2006): 167–198. Online at: tinyurl.com/7dsdfrr
> Tol, Richard S. J., and Gary W. Yohe. "A review of the Stern Review." *World Economics* 7:4 (2006): 233–250. Online at: tinyurl.com/826yrz4
>
> **Reports:**
> Goklany, I. M. "Death and death rates due to extreme weather events: Global and US trends 1900–2006." In *Civil Society Report on Climate Change*. London: International Policy Press, 2007; pp. 47–60. Online at: www.csccc.info/reports/report_20.pdf
> House of Lords. "The economics of climate change." HL Paper 12-1, Select Committee on Economic Affairs 2nd Report of Session 2005–06. London, 2005.
> Intergovernmental Panel on Climate Change, Working Group 1. "Climate change 2007: The physical science basis. Summary for policymakers." 4th Assessment Report (IPCC WG1 AR4 Report). 2007. Online at: tinyurl.com/88uagec [via archive.org].
> Reiter, Paul. "Human ecology and human behaviour: Climate change and health in perspective." In *Civil Society Report on Climate Change*. London: International Policy Press, 2007; pp. 21–46. Online at: www.csccc.info/reports/report_20.pdf

US Climate Change Science Program (CCSP). *Our Changing Planet: The US Climate Change Science Program for Fiscal Year 2009.*
Online at: www.climatescience.gov/infosheets/ccsp-8

Websites:

Global and Development Environment Institute at Tufts University: www.ase.tufts.edu/gdae

Intergovernmental Panel on Climate Change (IPCC): www.ipcc.ch

Science and Environmental Policy Project (SEPP): www.sepp.org

United Nations Environment Programme (UNEP) climate change pages: www.unep.org/climatechange/

US Climate Change Science Program (CCSP), integrating federal research on global change and climate change: www.climatescience.gov

Political Risk: Countering the Impact on Your Business
by Ian Bremmer
Eurasia Group, New York, USA

This Chapter Covers

- Business decision-makers must understand the political dynamics within the emerging market countries in which they operate.
- We can measure a state's stability—the ability of its government to implement policy and enforce laws despite a shock to the system.
- Essential to managing any type of risk is the development of a detailed and effective hedging strategy.
- Companies should not accept too much risk exposure within any one country or region.
- Rules of the game can change quickly in developing countries, and the cultivation of "friends in high places" isn't always a strong enough hedge.
- Operating in some developing countries comes with reputational risks at home.
- Too many companies have historically relied for insight into local politics and culture on employees who have lived in a particular country for only a short time—or have even merely traveled there.
- Those doing business in developing states need to have credible emergency response plans in place when events outside their control shut down supply chains, prevent local workers from coming to work, or otherwise disrupt operations.
- Developing strategies to recruit and train local managers serves several useful purposes.
- Devoting a share of profits to investment in local schools and universities, infrastructure, and charities can generate stores of goodwill, which is sometimes essential for cooperation with local workers and government officials.
- In some countries, foreign companies should be wary of transferring proprietary information to local partners or developing it inside the country.
- A foreign firm must look beyond what its local competitors are capable of producing today. It must anticipate how those capabilities are likely to develop over time.
- Conditions sometimes force companies to cut their losses and head for the exit. Ensuring that process is as painless and inexpensive as possible forms a crucial part of any sound risk-mitigation strategy.
- Political risk can be managed. It should not be avoided altogether.

Introduction
Over the past several years, and across a broad range of companies, corporate decision-makers seeking opportunities overseas have learned that it is not enough to have a knowledge of a foreign country's economic fundamentals. They also have to understand the forces and dynamics that shape these countries' politics. This is

especially true for emerging markets, where politics matters at least as much as economic factors for market outcomes. Of course, understanding that political risk matters is one thing. Knowing how to use it is another.

Stability

Starting with the basics, when committing a company to risk exposure in an emerging market country, it's essential to understand how political risk impacts the underlying strength of its government. There are two key elements to consider: stability and shock. Shocks are especially tough to forecast, because there are so many different kinds and because shocks are, by definition, unpredictable. We can't know when an earthquake will strike Pakistan, an elected leader will fall gravely ill in Nigeria, or a previously unknown group will carry out a successful terrorist attack in Indonesia.

But we can take the measure of a state's stability, which is defined as a government's ability to implement policy and enforce laws despite a shock to the system. The global financial crisis, a potent shock, has inflicted heavy losses on Russia's stock market. But Prime Minister Vladimir Putin has amassed plenty of political capital over the past several years, and President Dmitry Medvedev, his handpicked successor, basks in Putin's reflected glow. Neither need fear that large numbers of Russian citizens will turn on them anytime soon. In addition, a half-decade of windfall energy profits has generated more than US$500 billion in reserves, ready cash that can be used to bail out stock markets, banks, and, if necessary, an unpopular government. That's why, for the near-term, Russia will remain stable.

Pakistan is a different story. The country's newly elected government has a range of rivals and enemies. Inflation, power shortages, and a wave of suicide attacks have undermined the ruling Pakistan Peoples Party's domestic popularity. The financial crisis leaves the country at risk of debt default, forcing the government to negotiate a loan package with the International Monetary Fund that could impose austerity measures—the kind that helped topple civilian governments in Pakistan in the 1990s. The country is less stable than Russia, because it is much more vulnerable to the worst effects of shock.

President Luiz Inácio Lula da Silva has bolstered Brazil's stability over the past several years by quelling fears of left-wing populism with responsible (and predictable) macroeconomic policies. The Chinese Communist Party's ability to generate prosperity at home via three decades of successful economic liberalization has helped its leadership to build durable near-term stability.

But Nigeria's future stability remains at the mercy of President Umaru Yar'Adua's failing health, as historical tensions between northern Muslims and southern Christians combine with ongoing security challenges in the oil-rich Niger Delta region to prevent his government from building a national reputation for competence, vision, and strength. Iran's theocrats and firebrand president Mahmoud Ahmadinejad have effectively used the international conflict over the country's nuclear program to shore up support for the government in the face of high inflation and gasoline rationing. Underlying political factors in all these countries have a substantial impact on stability—and, therefore, on the country's business climate.

Political Risk: Countering the Impact on Your Business

Diversify
Yet it is not sufficient to possess broad insights into state stability. If corporate decision-makers are to design a credible business strategy that mitigates political risk and maximizes profit opportunities, they have to look deeper at the vulnerabilities that are peculiar to each country, each province, each community. Essential to managing any type of risk is the development of a detailed and effective diversification strategy. Given the political volatility within many developing world states—countries that will generate a large share of global growth over the next several decades—this kind of strategy is especially important. Even within a country as relatively stable as China, a closer look at internal political dynamics can identify various kinds of risk.

Two years ago, US officials worried publicly over a spike in sales of Russian arms to China. Dire predictions of a developing Russian–Chinese military axis became commonplace. But in 2007, sales of Russian arms to China fell by some 62%. Was it because the two governments had some sort of behind-the-scenes falling out? Did the Chinese leadership suddenly doubt the quality of Russian-made products? In reality, the arms sales slowed because China had mastered the design of many of the weapons, and Chinese companies began to produce them in sufficient quantities that demand for foreign-made weaponry fell sharply.

This is a cautionary tale, one that reminds us that any company betting heavily on long-term access to Chinese consumers (or to customers in many other developing countries) may be making a big mistake. There is plenty of money to be made in China for the next several years, but putting too many eggs in a single basket remains as risky as ever. For businesses with supply chains in China and other developing states, it's also important to build redundancies that are not overly exposed within any one region within these countries.

There are other, less obvious, components of a solid diversification strategy. Multinational companies should use all the leverage that their home governments and international institutions can provide to ensure that the governments of the countries in which they accept risk exposure protect their intellectual property rights, enforce all local laws intended to safeguard their commercial interests, and maintain open markets. Rules of the game can change quickly in developing countries, and the cultivation of "friends in high places" isn't always by itself an effective plan.

Know the Country
Gaining insight into a country's political, economic, social, and cultural traditions is essential for a successful risk-mitigation strategy. Where should this insight come from? Too many companies have historically relied on employees who have lived in a particular country for only a short time—or may even have done no more than travel there. Turning to the guy who backpacked through country X during college for useful information about its politics and culture—not as rare a phenomenon as you might think—is no substitute for the knowledge that can be gained from local workers themselves and from trained political risk analysts.

Design an Emergency Response
Generally speaking, emerging market countries are more vulnerable than rich world states to large-scale civil unrest, public health crises, and environmental disasters. Those doing business in developing states need credible emergency response plans in place when events outside their control shut down supply chains, prevent local workers from coming to work, or otherwise disrupt operations. Some businesses have designed technology plans that allow workers to work from home. In cases when circumstances force foreign workers to leave the country, locals should have the necessary training and skills to assume their responsibilities for an extended period. The added expense and time for training are well worth the cost. In some countries, they're essential.

Invest in Local Workers
Developing strategies to recruit and train local managers serves several useful purposes. First, it gives the host country government an investment in the success of a foreign-owned business. Every job created by a foreign firm is one that local government doesn't have to create. All governments want to keep unemployment at a minimum. Second, it gives local citizens a stake in the foreign company's success and helps to build solid relationships within the community. Some multinational firms have formed mutually profitable partnerships with local colleges and universities that give companies a fertile recruiting ground and ambitious students opportunities for work.

Invest in Their Communities
Devoting a share of profits to investment in local schools and universities, infrastructure, and charities can generate stores of goodwill, which is sometimes essential for cooperation with local workers and government officials. Yet, sensitivity to the local culture matters too. In many developing states, suspicions that Western (especially American) companies have a political or ideological agenda can undermine efforts to promote trust. Contributions to local quality of life should be seen to come without strings attached.

Protect Intellectual Property
In some countries, foreign companies should be wary of transferring proprietary information to local partners or developing it inside the country. Forging alliances with local partners in joint ventures often serves as an effective risk-mitigation strategy, but today's partner can become tomorrow's competitor, and a foreign firm can't always count on local courts or officials to safeguard its assets. Ironically, some foreign multinationals with long-term plans to remain inside a particular emerging market country have invested in local innovation. In the process, they have given locals an incentive to press their own government for stronger legal protections for intellectual property rights. Others have pooled their lobbying efforts with both local businesses and other foreign firms. When lobbying a government, strength in numbers can make a difference.

Know the Local Competition
Successful firms understand their comparative advantages. But a foreign company must look beyond what its local competitors are capable of producing today. It

Political Risk: Countering the Impact on Your Business

must anticipate how those capabilities are likely to develop over time. Identifying the markets in which a firm's core competencies are likely to deliver profits for the foreseeable future is essential for long-term risk-mitigation strategies.

In many emerging market countries, local companies are often better at large-scale efficient manufacturing than at designing products, marketing them, and delivering them to the customer. Knowing how quickly the local competition can climb the value chain helps with the design of an intelligent, long-term business strategy.

Know Where to Find the Exits

Many companies have made lots of money in emerging markets. But as Wall Street veterans like to say, "Don't confuse brilliance with a bull market." Some companies have gotten away with ignoring the need for solid risk-management strategies and have simply ridden the wave produced by the inevitable rise of emerging market economies.

Yet, as skepticism of globalization grows in some developing countries, as their governments respond to domestic political pressure by rewriting rules to favor local companies at the expense of their foreign competitors, and as the challenges facing multinational companies operating inside these countries become more complex, it's important to have an exit strategy. There are plenty of developing states that are now open for business and investment. They have different strengths and vulnerabilities. Too much risk exposure in any one of them can create unnecessary risks. Conditions sometimes force companies to cut their losses and head for the door. Ensuring that this process is as painless and inexpensive as possible forms a crucial part of any sound risk-mitigation strategy.

Don't Forget the Power of Perception

Operating in some developing countries comes with reputational risks at home. Several US companies have faced tough domestic criticism for doing business with governments that are accused of violating international labor, environmental, and human-rights standards. For a company's leadership, clearly communicating what the company will and won't do to gain market access in certain countries—and strict adherence to these standards of conduct—can help to minimize this risk.

Political Risk Insurance

As a last resort, a firm can purchase political risk insurance from providers like the Multilateral Investment Guarantee Agency, an arm of the World Bank, or the US government's Overseas Private Investment Corporation. But this should be a last resort strategy, because high premiums, substantial transaction and opportunity costs, and the complexities of establishing a valid claim have taught many companies that it is far more cost-effective to prevent or pre-empt bad outcomes than to rely heavily on plans to cope with their aftermath.

A Little Tolerance Is a Good Thing

It's useful to remember that having a good exit strategy does not require you to use it. Doing business in developing states comes with risk. But refusing to enter these markets or pulling out at the first sign of trouble comes with a high cost to opportunity.

Risk Management in an Uncertain World

Foreign companies will be earning solid profits within emerging market states for many years to come. Political risk can be managed. It should not be avoided altogether.

More Info

Books:

Bracken, Paul, Ian Bremmer, and David Gordon (eds). *Managing Strategic Surprise: Lessons from Risk Management and Risk Assessment.* New York: Cambridge University Press, 2008.

Howell, Llewellyn D. (ed). *Handbook of Country and Political Risk Analysis.* 3rd ed. East Syracuse, NY: Political Risk Services Group, 2002.

Moran, Theodore H. (ed). *Managing International Political Risk.* London: Blackwell Publishing, 1999.

Moran, Theodore H., Gerald T. West, and Keith Martin (eds). *International Political Risk Management: Meeting the Needs of the Present, Anticipating the Challenges of the Future.* Washington, DC: World Bank Publications, 2007.

Wilkin, Sam (ed). *Country and Political Risk: Practical Insights for Global Finance.* London: Risk Books, 2004.

Articles:

Bremmer, Ian, and Fareed Zakaria. "Hedging political risk in China." *Harvard Business Review* 84:11 (November 2006): 22–25. Online at: tinyurl.com/cln9l5j

The Economist. "Insuring against political risk." April 4, 2007. Online at: tinyurl.com/csewtgb

Henisz, Witold J., and Bennet A. Zelner. "Political risk management: A strategic perspective." Online at: tinyurl.com/89rekyj [PDF].

Stanislav, Markus. "Corporate governance as political insurance: Firm-level institutional creation in emerging markets and beyond." *Socio-Economic Review* 6:1 (January 2008): 69–98. Online at: dx.doi.org/10.1093/ser/mwl036

Report:

PricewaterhouseCoopers. "Integrating political risk into enterprise risk management." 2006. Online at: tinyurl.com/ksjp7g

Websites:

Eurasia Group, global political risk advisory and consulting firm: www.eurasiagroup.net

Multilateral Investment Guarantee Agency (MIGA)'s Political Risk Insurance Center: www.pri-center.com

PricewaterhouseCoopers: www.pwc.com. Enter "political risk" in search box to find articles and resources.